PRACTICE PROBLEMS WORKBOOK

ENGINEERING MECHANICS
S T A T I C S
TENTH EDITION

R. C. Hibbeler

PEARSON

Prentice Hall

Upper Saddle River, New Jersey 07458

Executive Editor: *Eric Svendsen*
Associate Editor: *Dee Bernhard*
Supplement Cover Manager: *Daniel Sandin*
Executive Managing Editor: *Vince O'Brien*
Managing Editor: *David A. George*
Production Editor: *Barbara A. Till*
Buyer: *Ilene Kahn*

© 2004, 2001, 1998, 1995, 1992, 1989, 1986, 1983, 1978, 1974 by
R. C. Hibbeler
Published by Pearson Prentice Hall
Pearson Education, Inc.
Upper Saddle River, NJ 07458

Printed in the United States of America
10 9 8 7 6 5 4 3 2 1

ISBN 0-13-141211-6

Pearson Education Ltd., *London*
Pearson Education Australia Pty. Ltd., *Sydney*
Pearson Education Singapore, Pte. Ltd.
Pearson Education North Asia Ltd., *Hong Kong*
Pearson Education Canada, Inc., *Toronto*
Pearson Educación de Mexico, S.A. de C.V.
Pearson Education—Japan, *Tokyo*
Pearson Education Malaysia, Pte. Ltd.
Pearson Education, Inc., *Upper Saddle River, New Jersey*

Preface

This workbook is a supplement to the textbook *Engineering Mechanics: Statics*. As a result, the problems in this book are arranged in the same order as those presented in the textbook. Here the solution to the problems is only partially complete. The key equations, which stress the important fundamentals of the problem solution, must be supplied in the space provided. There is no need for calculations, however, since all the answers are given in the back of the book.

It is suggested that these problems by solved just after the theory and example problems covering the corresponding topic have been studied in the textbook. If an honest effort is made at completing and understanding the solution to these problems, it will serve to build confidence in applying the theory to the textbook problems. Furthermore, these problems provide an excellent review of the subject matter, which can then be used when preparing for exams.

I would greatly appreciate hearing from you if you have any comments or suggestions regarding the contents of this work.

Russell Charles Hibbeler
hibbeler@bellsouth.net

Contents

1 General Mathematical Principles

The mathematical concepts provided by this review should be thoroughly mastered for success in solving problems in mechanics.

Solution of Linear Algebraic Equations

Such a set of equations may be solved by sucessive elimination of the unknown variables, until one is left with (having one unknown. After solving for this unknown, the other unknowns are determined by the process of back substitution. The following example, involving three equations with three unknowns (x,y,z) illustrates the procedure.

$$-x + 4y + z = 1 \tag{1}$$
$$2x - y + z = 2 \tag{2}$$
$$4x - 5y + 3z = 4 \tag{3}$$

Eliminate one of the variables, for example solve for x in Eq. (1) and Eq. (2), both in terms of y and z. This yields,

$$x = 4y + z - 1 \tag{4}$$
$$x = \frac{1}{2}y - \frac{1}{2}z + 1 \tag{5}$$

Substitute each of these equations into Eq. (3) and simplify, which gives,

$$11y + 7z = 8 \tag{6}$$
$$-3y + z = 0 \tag{7}$$

We now have eliminated x and have two equations with two unknowns. Solve Eq. (7) for y in terms of z, substitute into Eq. (6) and solve for z; i.e.,

$$y = \frac{1}{3}z \tag{8}$$

So that,

$$\frac{11}{3}z + 7z = 8$$

Therefore,

$$z = 0.750 \qquad \textbf{Ans.}$$

Now by back substituion of z solve for y using Eq. (8),

$$y = \frac{1}{3}(0.750) = 0.250 \qquad \textbf{Ans.}$$

then for x, using Eq. (4),

$$x = 4(0.250) + 0.750 - 1 = 0.750 \qquad \textbf{Ans.}$$

The solution may be rechecked by substituting the answers into the original three equations.

Solution of a Quadratic Equation

The standard form of this equation is

$$ax^2 + bx + c = 0$$

The solution is determined from the quadratic formula

$$x = \frac{-b \pm \sqrt{b^2 - 4ac}}{2a}$$

For example, to solve $x^2 + 2x = 4$, the equation is rearranged in the standard form

$$x^2 + 2x - 4 = 0$$

So that $a = 1$, $b = 2$, $c = -4$. Thus,

$$x = \frac{-2 \pm \sqrt{2^2 - 4(1)(-4)}}{2(1)} = \frac{-2 \pm 4.47}{2}$$

$$x = 1.24 \text{ and } x = -3.24$$

Ans.

Constructed Relationships Between Angles

Note how the angles between parallel lines and sloping and normal lines are related.

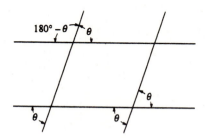

For the cases involving sloping and normal lines, the angles are easy to establish if one imagines $\theta \to 0$, so that the incline lines approach the horizontal or vertical coincidently.

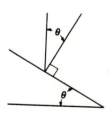

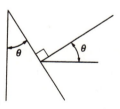

2

All *triangles* inscribed in a semicircle have a *right angle*.

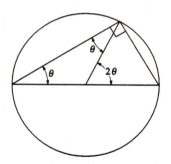

Right Triangles

$$C = \sqrt{A^2 + B^2} \qquad \text{(Pythagorean Theorem)}$$
$$a + b = 90°$$
$$\text{Area} = \frac{1}{2}AB$$

In particular, note the 3 - 4 - 5 and 5 - 12 - 13 right triangles :

If two triangles are similar, their sides are respectively proportional. For example, if ABC is similar to $A'B'C'$, then

$$\frac{AC}{AB} = \frac{A'C'}{A'B'} \text{ or } \frac{AC}{6} = \frac{5}{3}, AC = 10$$
$$\frac{CB}{AB} = \frac{C'B'}{A'B'} \text{ or } \frac{CB}{6} = \frac{4}{3}, CB = 8$$

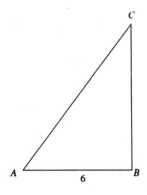

3

Circle

Arc length is defined by $s = \theta r$, where θ is in radians and s and r have the same dimensions. $360° = 2\pi$ radians, so that 1 radian $= 180°/\pi$.

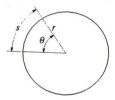

Right triangles

For a given right triangle we define the basic trigonometric functions as

$$\sin (angle) = \frac{\text{side opposite to the angle}}{\text{hypotenuse}}$$

$$\cos (angle) = \frac{\text{side adjacent to the angle}}{\text{hypotenuse}}$$

$$\tan (angle) = \frac{\text{side opposite to the angle}}{\text{side adjacent to the angle}}$$

These equations are easily remembered if one memorizes the acronyn, SOH - CAH - TOA, where S, C, and T are sine, cosine, and tangent, respectively, and O is the opposite side, A is the adjacent side, and H is the hypotenuse. For example, given the right triangle shown in the figure,

$$\sin \theta = \frac{4}{5} \qquad \cos \theta = \frac{3}{5} \qquad \tan \theta = \frac{4}{3}$$

The angle θ can be determined from a pocket calculator. Mathematically, we designate this by the notation

$$\theta = \sin^{-1} \frac{4}{5} \qquad \theta = \cos^{-1} \frac{3}{5} \qquad \theta = \tan^{-1} \frac{4}{3}$$

where $\sin^{-1}$ denotes the arc or inverse sine etc. of the angle.

In a similar manner, given any right triangle, we can determine the sides AB and BC from the triangles shown by using a pocket calculator.

$$\cos 20° = \frac{AB}{3}; \ AB = 3 \cos 20° = 2.82$$

$$\sin 20° = \frac{BC}{3}; \ BC = 3 \sin 20° = 1.03$$

In particular, from the special triangles shown note the following :

$$\sin 45° = \frac{1}{\sqrt{2}} = 0.7071 \qquad \sin 30° = \cos 60° = \frac{1}{2} = 0.5$$

$$\cos 45° = \frac{1}{\sqrt{2}} = 0.7071 \qquad \sin 60° = \cos 30° = \frac{\sqrt{3}}{2} = 0.8660$$

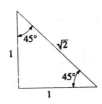 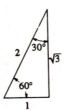

Nonright triangles

The sides and angles of a nonright triangle are determined from

Sine law

$$\frac{A}{\sin a} = \frac{B}{\sin b} = \frac{C}{\sin c}$$

Cosine law

$$C = \sqrt{A^2 + B^2 - 2AB \cos c}$$

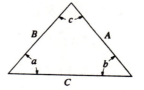

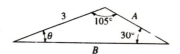

For example, for the triangle shown in the figure, the unknown side B can be determined as follows :

$$\frac{B}{\sin 105°} = \frac{3}{\sin 30°}, \qquad B = \frac{3(0.9659)}{(0.5)} = 5.80$$

Since $\theta = 180° - 105° - 30° = 45°$, A may also be obtained by the law of sines

$$\frac{A}{\sin 45°} = \frac{3}{\sin 30°}, \qquad A = \frac{3(0.7071)}{0.5} = 4.24$$

or, using the results for B and θ, we can obtain A by the law of cosines, i.e.,

$$A = \sqrt{(3)^2 + (5.80)^2 - 2(3)(5.80) \cos 45°} = 4.24$$

Problems

1.1 Solve the following three equations for x, y, and z :

$$x - y + z = -1 \qquad -x + y + z = -1 \qquad x + 2y - 2z = 5$$

1.2 In each case, solve the following quadratic equations for the two roots of x :

a) $x^2 - 16 = 0$ b) $-x^2 + 5x = -6$

1.3 In each case, determine the angle θ :

a)   b)   c)

  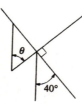

1.4 Determine the length of side AB if $\triangle ABC$ is similar to $\triangle A'B'C'$.

 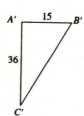

1.5 In each case, using the basic trigonometric functions, determine the length of side AB :

a)   b)

 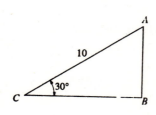 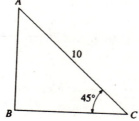

c)

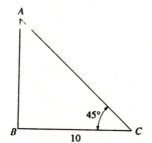

1.6 Determine the angles ϕ and θ, and the side AB of the triangle :

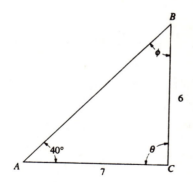

1.7 Using the law of sines or cosines determine the length of the unknown sides in each triangle.

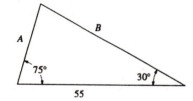 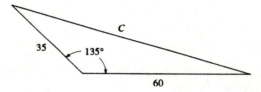

2 Force Vectors

Vector Addition of Forces Using the Parallelogram Law

2 - 1. Determine the magnitude of the resultant force and its direction measured clockwis*e* from the positive *x* axis.

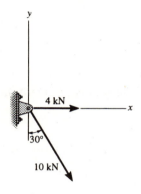

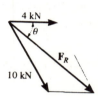

Solution

The parallelogram is first drawn to show the vector addition. From this we have the triangle, tip - to - tail costruction. Apply the law of cosines to find F_R

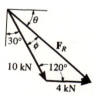

$F_R = $ _____

$F_R = 12.5$ kN   **Ans.**

Using this result, apply the law of sines to find ϕ

_____

$\phi = 16.10°$

$\theta = 90° - 16.10° - 30° = 43.9°$ **Ans.**

2 - 2. Determine the magnitude and direction θ of **F** so that this force has components of 40 N acting from A toward B and 60 N acting from A toward C on the frame.

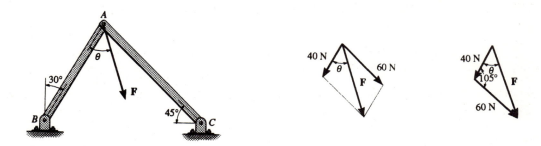

Solution

The resultant **F** is resolved into the 40 N and 60 N component forces formed by extending parallel lines from the tip of **F** as shown.

Noting that the interior angle is $180° - 30° - 45° = 105°$, apply the law of cosines to determine F.

$F =$ _____

$F = 80.3$ N   **Ans.**

Use this result and apply the law of sines to determine θ.

_____

$\theta = 46.2°$ **Ans.**

2 - 3. Determine the design angle θ ($\theta < 90°$) between the two struts so that the 500 - N horizontal force has a component of 600 N directed from A toward C. What is the component of force acting along member AB?

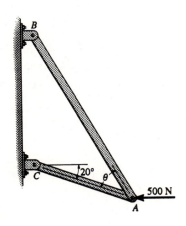

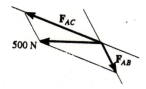

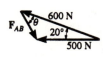

Solution

First the 500 - N force is resolved into its components by drawing lines parallel to AB and AC from the tip of the force as shown.

Using the triangle, apply the law of cosines to determine F_{AB}.

$$F_{AB} = \text{\underline{\hspace{10cm}}}$$

$$F_{AB} = 214.9 \text{ N} = 215 \text{ N} \qquad\qquad \textbf{Ans.}$$

Use this result and apply the law of sines to determine θ.

$$\text{\underline{\hspace{8cm}}}$$

$$\theta = 52.7° \qquad\qquad \textbf{Ans.}$$

2 - 4. Determine the magnitudes of $\mathbf{F}_1$ and $\mathbf{F}_2$ so that the resultant force has a magnitude of 20 N and is directed along the positive x axis.

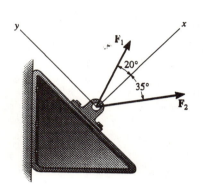 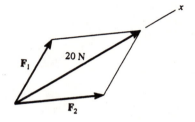 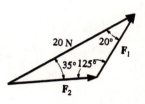

Solution

The 20 - N resultant force is resolved into components $\mathbf{F}_1$ and $\mathbf{F}_2$ by drawing lines parrallel to $\mathbf{F}_1$ and $\mathbf{F}_2$ from the tip of the resultant as shown . Using the triangle, apply the law of sines to obtain F_1 and F_2.

$$F_1 = 14.0 \text{ N}$$ **Ans.**

$$F_2 = 8.35 \text{ N}$$ **Ans.**

2 - 5. The plate is subjected to the two member forces at A and B. Determine the magnitude and direction of the resultant force.

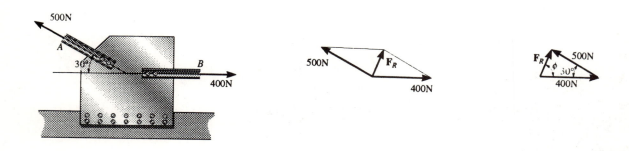

Solution

The resultant is formed by extending parallel lines from the tip of the 500 N and 400 N forces as shown. Using the triangle, apply the law of cosines to determine F_R.

$$F_R = \underline{\hspace{8cm}}$$

$$F_R = 252.2 \text{ N} = 252 \text{ N} \qquad\qquad \textbf{Ans.}$$

Using this result, apply the law of sines to determine ϕ.

$$\underline{\hspace{8cm}}$$

$$\phi = 82.5° \qquad\qquad \textbf{Ans.}$$

2 - 6. Determine the required angle θ for connecting member A to the plate if the resultant of the two forces is to be directed vertically upward. Also, what is the magnitude of the resultant?

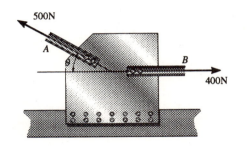 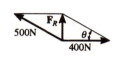

Solution

The resultant is resolved into components along A and B by extending lines parallel to A and B from the tip of F_R as shown. Using the triangle, determine the required angle θ.

_____

$\theta = 36.9°$ **Ans.**

Write an equation to determine the resultant force.

$F_R =$ _____

$F_R = 300 \text{ N}$ **Ans.**

Addition of Rectangular Force Components

2 - 7. Determine the x and y components of the 2 - kN force.

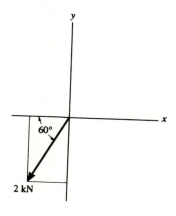

2 kN

Solution

Resolve the force into its rectangular components. Here

$F_x =$ _____ $= -1.00$ kN $= 1.00$ kN $\leftarrow$ **Ans.**

$F_y =$ _____ $= -1.73$ kN $= 1.73$ kN $\downarrow$ **Ans.**

2 - 8. Resolve the 700 - N force into its x and y components and determine these components.

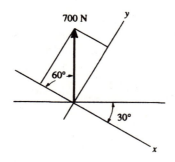

Solution

The components are

$$F_x = \underline{\hspace{5cm}} = 350 \text{ N} \qquad \textbf{Ans.}$$

$$F_y = \underline{\hspace{5cm}} = -606 \text{ N} = 606 \text{ N} \qquad \textbf{Ans.}$$

2 - 9. Determine the magnitude and direction of the resultant force by adding the rectangular components of the three forces.

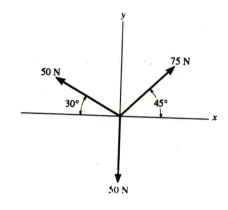

Solution

$\xrightarrow{+} F_{R_x} = \Sigma F_x;$ $F_{R_x} = \underline{\hspace{6cm}}$

$$F_{R_x} = 9.732 \text{ N} \rightarrow$$

$+\uparrow F_{R_y} = \Sigma F_y;$ $F_{R_y} = \underline{\hspace{6cm}}$

$$F_{R_y} = 28.03 \text{ N} \uparrow$$

$$F_R = \sqrt{(9.732)^2 + (28.03)^2} = 29.7 \text{ N}$$ **Ans.**

$$\theta = \tan^{-1} = \frac{28.03}{9.732} = 70.9° \angle$$ **Ans.**

2 - 10. Determine the magnitude and direction of the resultant force by adding the rectangular components.

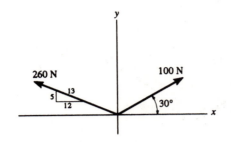

Solution

$\overset{+}{\rightarrow} F_{R_x} = \Sigma F_x;$ $F_{R_x} = $ _____

$$F_{R_x} = -153.4\,N = 153.4\,N \leftarrow$$

$+ \uparrow F_{R_y} = \Sigma F_y;$ $F_{R_y} = $ _____

$$F_{R_y} = 150\,N \uparrow$$

$$F_R = \sqrt{(153.4)^2 + (150)^2} = 215\,N \qquad\qquad \textbf{Ans.}$$

$$\theta = \tan^{-1}\frac{150}{153.4} = 44.4^\circ \qquad\qquad \textbf{Ans.}$$

2 - 11. Determine the magnitude and direction of the resultant force by adding the rectangular componets.

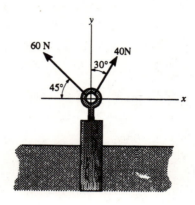

Solution

$\overset{+}{\to} F_{R_x} = \Sigma F_x;$ $F_{R_x} =$ _____

$$F_{R_x} = -22.43 \text{ N} = 22.43 \text{ N} \leftarrow$$

$+\uparrow F_{R_y} = \Sigma F_y;$ $F_{R_y} =$ _____

$$F_{R_y} = 77.07 \text{ N} \uparrow$$

$$F_R = \sqrt{(-22.43)^2 + (77.07)^2} = 80.3 \text{ N} \qquad \textbf{Ans.}$$

$$\theta = \tan^{-1}\frac{(77.07)}{(22.43)} = 73.8° \qquad \textbf{Ans.}$$

2 - 12. Determine the angles θ and ϕ so that the resultant force is directed along the positive x axis and has a magnitude $F_R = 20$ N.

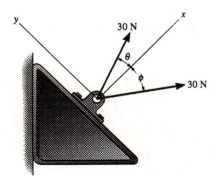

Solution

$+ \; F_{R_x} = \Sigma F_x ; \qquad \underline{\hspace{2cm}} = \underline{\hspace{6cm}}$

$+ \; F_{R_y} = \Sigma F_y ; \qquad \underline{\hspace{2cm}} = \underline{\hspace{6cm}}$

Solving,

$$\phi = \theta = 70.5° \qquad \qquad \textbf{Ans.}$$

Force Expressed as a Cartesian Vector

2 - 13. Express each force in Cartesian vector form.

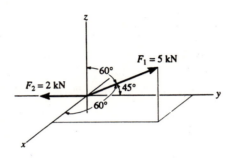

Solution

Use the direction cosines to establish the x, y, z, components of $\mathbf{F}_1$.

$$\mathbf{F}_1 = \underline{\hspace{6cm}}$$

$$\mathbf{F}_1 = \{2.50\mathbf{i} + 3.54\mathbf{j} + 2.50\mathbf{k}\} \text{ kN} \qquad \textbf{Ans.}$$

$$\mathbf{F}_2 = \underline{\hspace{4cm}} \qquad \textbf{Ans.}$$

2 - 14. Express each force in Cartesian vector form.

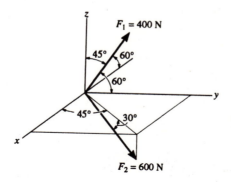

Solution

Note for $\mathbf{F}_1$ that $\alpha_1 = 180° - 60° = 120°$.

$$\mathbf{F}_1 = \underline{\hspace{10cm}}$$

$\mathbf{F}_1 = \{-200\mathbf{i} + 200\mathbf{j} + 283\mathbf{k}\}$ N   **Ans.**

Note that the angles 45° and 30° defining the direction of $\mathbf{F}_2$ are not coordinate direction angles for $\mathbf{F}_2$.

$$\mathbf{F}_2 = \underline{\hspace{10cm}}$$

$\mathbf{F}_2 = \{367\mathbf{i} + 367\mathbf{j} - 300\mathbf{k}\}$ N   **Ans.**

2 - 15. The ball joint is subjected to the three forces shown. Express each force in Cartesian vector form and then determine the magnitude and coordinate direction angles of the resultant force.

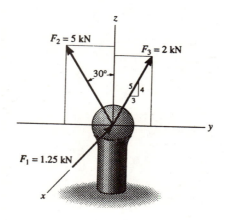

Solution

$\mathbf{F}_1 = $ _____

$\mathbf{F}_2 = $ _____

$\mathbf{F}_2 = \{-2.50\mathbf{j} + 4.330\mathbf{k}\}$ kN

$\mathbf{F}_3 = $ _____

$\mathbf{F}_3 = \{1.20\mathbf{j} + 1.60\mathbf{k}\}$ kN

$\mathbf{F}_R = \Sigma\mathbf{F} = \{-1.25\mathbf{i} - 1.30\mathbf{j} + 5.930\mathbf{k}\}$ kN

$F_R = $ _____

$F_R = 6.198$ kN $= 6.20$ kN   **Ans.**

$\alpha = \cos^{-1}(\text{———}) = 102°$ **Ans.**

$\beta = \cos^{-1}(\text{———}) = 102°$ **Ans.**

$\gamma = \cos^{-1}(\text{———}) = 16.9°$ **Ans.**

22

2-16. Determine the magnitude and coordinate direction angles of $\mathbf{F}_2$ so that the resultant of the two forces acts upward along the z axis of the pole and has a magnitude of 275 N.

Solution

For $\mathbf{F}_1$, note that $\alpha_1 = 180° - 45° = 135°$.

$$\mathbf{F}_1 = \underline{\hspace{5cm}}$$

$$= \{-141.4\mathbf{i} + 100.0\mathbf{j} + 100.0\mathbf{k}\} \text{ N}$$

$$\mathbf{F}_2 = F_{2x}\mathbf{i} + F_{2y}\mathbf{j} + F_{2z}\mathbf{k}$$

$$\mathbf{F}_R = \underline{\hspace{5cm}}$$

$$F_{R_x} = \Sigma F_x; \underline{\hspace{1.5cm}} = \underline{\hspace{4cm}}$$

$$F_{2x} = 141.4 \text{ N}$$

$$F_{R_y} = \Sigma F_y; \underline{\hspace{1.5cm}} = \underline{\hspace{4cm}}$$

$$F_{2y} = -100 \text{ N}$$

$$F_{R_z} = \Sigma F_z; \underline{\hspace{1.5cm}} = \underline{\hspace{4cm}}$$

$$F_{2z} = 175 \text{ N}$$

$$F_2 = \sqrt{(141.4)^2 + (-100)^2 + (175)^2} = 246.2 \text{ N} = 246 \text{ N} \qquad \textbf{Ans.}$$

$$\alpha = \cos^{-1}(\frac{141.4}{246.2}) = 54.9° \qquad \textbf{Ans.}$$

$$\beta = \cos^{-1}(\frac{-100}{246.2}) = 114° \qquad \textbf{Ans.}$$

$$\gamma = \cos^{-1}(\frac{175}{246.2}) = 44.7° \qquad \textbf{Ans.}$$

2 - 17. Force **F** acts on peg A such that one of its components, lying in the $x - y$ plane, has a magnitude of 50 N as shown. Express **F** as a Cartesian vector.

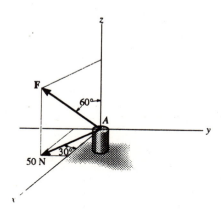

Solution

$$F_x = \underline{\hspace{8cm}} = 43.3 \text{ N}$$

$$F_y = \underline{\hspace{8cm}} = -25.0 \text{ N}$$

$$F = \underline{\hspace{8cm}} = 57.74 \text{ N}$$

$$F_z = \underline{\hspace{8cm}} = 28.9 \text{ N}$$

Thus,

$$\mathbf{F} = \{43.3\mathbf{i} - 25.0\mathbf{j} + 28.9\mathbf{k}\} \text{ N} \qquad\qquad \textbf{Ans.}$$

2 - 18. Specify the magnitude and cordinate direction angles α_1, β_1, γ_1 of $\mathbf{F}_1$ so that the resultant of the three forces acting on the post is $\mathbf{F}_R = \{-350\mathbf{k}\}$ N. Note that $\mathbf{F}_3$ lies in the $x - y$ plane.

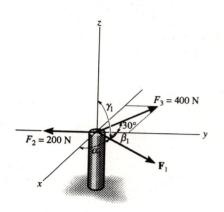

Solution

$$\mathbf{F}_1 = F_{1x}\mathbf{i} + F_{1y}\mathbf{j} + F_{1z}\mathbf{k}$$

$$\mathbf{F}_2 = \underline{\hspace{5cm}}$$

$$\mathbf{F}_3 = \underline{\hspace{5cm}}$$

$$= \{-200\mathbf{i} + 346.4\mathbf{j}\}\ \text{N}$$

$$\mathbf{F}_R = \Sigma\mathbf{F}$$

$$-350\mathbf{k} = F_{1x}\mathbf{i} + F_{1y}\mathbf{j} + F_{1z}\mathbf{k} - 200\mathbf{j} - 200\mathbf{i} + 346.4\mathbf{j}$$

$$F_{1x} = 200\ \text{N}$$

$$F_{1y} = 200 - 346.4 = -146.4\ \text{N}$$

$$F_{1z} = -350\ \text{N}$$

$$F_1 = \sqrt{(200)^2 + (-146.4)^2 + (-350)^2} = 428.9\ \text{N} = \ 429\ \text{N} \qquad\qquad \textbf{Ans.}$$

$$\alpha_1 = \cos^{-1}(\frac{200}{428.9}) = 62.2° \qquad\qquad \textbf{Ans.}$$

$$\beta_1 = \cos^{-1}(\frac{-146.4}{428.9}) = 110° \qquad\qquad \textbf{Ans.}$$

$$\gamma_1 = \cos^{-1}(\frac{-350}{428.9}) = 145° \qquad\qquad \textbf{Ans.}$$

Position Vectors

2 - 19. Express the position vector **r** in Cartesian vector form; then determine its magnitude and coordinate direction angles.

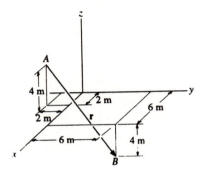

Solution

The easiest way to establish **r** is to think of the distance one must travel along the x, y, z axes to go from A to B.

r = _____

r = _____ = 12 m   **Ans.**

$\alpha = \cos^{-1}(\underline{\quad\quad}) = 70.5°$ **Ans.**

$\beta = \cos^{-1}(\underline{\quad\quad}) = 48.2°$ **Ans.**

$\gamma = \cos^{-1}(\underline{\quad\quad}) = 132°$ **Ans.**

2 - 20. The cord is attached between two walls. If it is 8 m long, determine the distance x to the point of attachment at B.

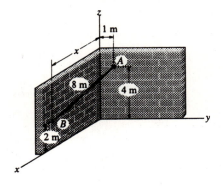

Solution

Determine the position vector directed from A to B.

$$\mathbf{r}_{AB} = \underline{\hspace{10cm}}$$

Magnitude

$$8^2 = x^2 + (-1)^2 + (-2)^2$$

$$x = 7.68 \text{ m} \hspace{6cm} \textbf{Ans.}$$

Force Vector Directed Along a Line

2 - 21. Express **F** as a Cartesian vector; then determine its coordinate direction angles.

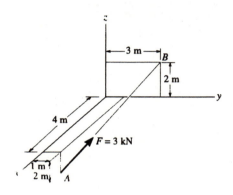

Solution

$$\mathbf{r}_{AB} = \underline{\hspace{5cm}}$$

$$r_{AB} = \sqrt{(-4)^2 + (2)^2 + (4)^2} = 6 \text{ m}$$

$$\mathbf{F} = \underline{\hspace{5cm}}$$

$$\mathbf{F} = \{-2\mathbf{i} + 1\mathbf{j} + 2\mathbf{k}\} \text{ kN} \qquad\qquad \textbf{Ans.}$$

$$\alpha = \cos^{-1}(\frac{-2}{3}) = 132° \qquad\qquad \textbf{Ans.}$$

$$\beta = \cos^{-1}(\frac{1}{3}) = 70.5° \qquad\qquad \textbf{Ans.}$$

$$\gamma = \cos^{-1}(\frac{2}{3}) = 48.2° \qquad\qquad \textbf{Ans.}$$

2 - 22. The antenna tower is supported by three cables. If the forces in these cables are $F_B = 520$ N, $F_C = 680$ N, and $F_D = 560$ N, determine the magnitude and coordinate direction angles of the resultant force acting at A.

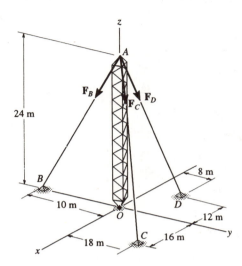

Solution

$$\mathbf{F}_B = F_B\left(\frac{\mathbf{r}_{AB}}{r_{AB}}\right) = \underline{\hspace{4cm}} = \{-200\mathbf{j} - 480\mathbf{k}\} \text{ N}$$

$$\mathbf{F}_C = F_C\left(\frac{\mathbf{r}_{AC}}{r_{AC}}\right) = \underline{\hspace{4cm}} = \{320\mathbf{i} + 360\mathbf{j} - 480\mathbf{k}\} \text{ N}$$

$$\mathbf{F}_D = F_D\left(\frac{\mathbf{r}_{AD}}{r_{AD}}\right) = \underline{\hspace{4cm}} = \{-240\mathbf{i} + 160\mathbf{j} - 480\mathbf{k}\} \text{ N}$$

$$\mathbf{F}_R = \Sigma\mathbf{F} = \{80\mathbf{i} + 320\mathbf{j} - 1440\mathbf{k}\} \text{ N}$$

$$F_R = \sqrt{(80)^2 + (320)^2 + (-1440)^2} = 1477.3 \text{ N} = 1.48 \text{ kN} \qquad \textbf{Ans.}$$

$$\alpha = \cos^{-1}\left(\frac{80}{1477.3}\right) = 86.9° \qquad \textbf{Ans.}$$

$$\beta = \cos^{-1}\left(\frac{320}{1477.3}\right) = 77.5° \qquad \textbf{Ans.}$$

$$\gamma = \cos^{-1}\left(\frac{-1440}{1477.3}\right) = 167° \qquad \textbf{Ans.}$$

2 - 23. The cable AO exerts a force on the top of the pole of $\mathbf{F} = \{-120\mathbf{i} - 90\mathbf{j} - 80\mathbf{k}\}$ N. If the cable has a length of 34 m, determine the height z of the pole and the location (x, y) of its base.

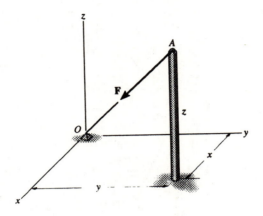

Solution

The magnitude of $\mathbf{F}$ is

$$F = \underline{\hspace{6cm}} = 170 \text{ N}$$

Thus

$$\mathbf{u}_{AO} = (\underline{\hspace{1cm}})\mathbf{i} + (\underline{\hspace{1cm}})\mathbf{j} + (\underline{\hspace{1cm}})\mathbf{k}$$

$$\mathbf{r}_{AO} = 34\mathbf{u}_{AO} = \{-24\mathbf{i} - 18\mathbf{j} - 16\mathbf{k}\} \text{ m}$$

$x = 24 \text{ m}$ **Ans.**

$y = 18 \text{ m}$ **Ans.**

$z = 16 \text{ m}$ **Ans.**

Dot Product

2 - 24. Determine the angle θ between the axis of the pole and the wire AB.

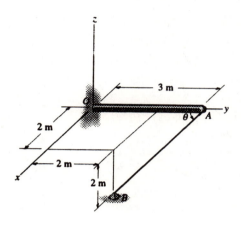

Solution

$$\mathbf{r}_{AO} = \underline{\hspace{6cm}}$$

$$r_{AO} = 3 \text{ m}$$

$$\mathbf{r}_{AB} = \underline{\hspace{6cm}}$$

$$r_{AB} = 3 \text{ m}$$

$$\mathbf{r}_{AO} \cdot \mathbf{r}_{AB} = \underline{\hspace{7cm}}$$

$$\theta = \cos^{-1}(\underline{\hspace{2cm}}) = 70.5° \qquad \textbf{Ans.}$$

2 - 25. Determine the projection of the position vector **r** along the *ab* axis.

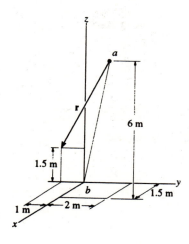

Solution

$\mathbf{r} =$ _____

$$\mathbf{u}_{ab} = (\frac{}{6.5})\mathbf{i} + (\frac{}{6.5})\mathbf{j} + (\frac{}{6.5})\mathbf{k}$$

$$r_p = \mathbf{r} \cdot \mathbf{u}_{ab} = (\quad)(\quad) + (\quad)(\quad) + (\quad)(\quad)$$

$$r_p = 5.42 \text{ m} \qquad\qquad\qquad\qquad\qquad \textbf{Ans.}$$

2 - 26. Cable BC exerts a force of $F = 28$ N on the top of the flagpole. Determine the projection of this force along the z axis of the pole.

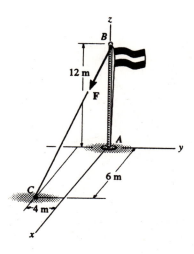

Solution

$$r_{BC} = \underline{\hspace{1cm}} i + \underline{\hspace{1cm}} j + \underline{\hspace{1cm}} k$$

$$r_{BC} = 14 \text{ m}$$

$$F = 28[(\underline{\hspace{1.5cm}})i + (\underline{\hspace{1.5cm}})j + (\underline{\hspace{1.5cm}})k]$$

$$F = \{12i - 8j - 24k\} \text{ N}$$

$$u = \underline{\hspace{6cm}}$$

$$F_P = F \cdot u = \underline{\hspace{8cm}}$$

$$F_P = 24 \text{ N} \qquad\qquad\qquad\qquad\qquad \textbf{Ans.}$$

2 - 27. Two forces act on the hook. Determine the angle θ between them. Also, what is the projection of $\mathbf{F}_2$ along the y axis?

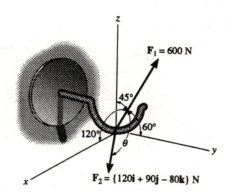

F$_1$ = 600 N

45°

120°

60°

θ

z

y

x

$\mathbf{F}_2 = \{120\mathbf{i} + 90\mathbf{j} - 80\mathbf{k}\}$ N

Solution

Express $\mathbf{F}_1$ as a Cartesian vector.

$$\mathbf{F}_1 = \underline{\hspace{5cm}}$$

$$= \{-300\mathbf{i} + 300\mathbf{j} + 424.3\mathbf{k}\} \text{ N}$$

$$\mathbf{F}_2 = \{120\mathbf{i} + 90\mathbf{j} - 80\mathbf{k}\} \text{ N}$$

$$F_2 = 170 \text{ N}$$

$$\mathbf{F}_1 \cdot \mathbf{F}_2 = \underline{\hspace{5cm}}$$

$$= -42\,941 \text{ N}^2$$

$$\theta = \cos^{-1}(\underline{\hspace{2cm}}) = 115° \qquad \textbf{Ans.}$$

The projection of $\mathbf{F}_2$ along the y axis is

$$F_{P_2} = \mathbf{F}_2 \cdot \mathbf{u} = \underline{\hspace{5cm}}$$

$$F_{P_2} = 90 \text{ N} \qquad \textbf{Ans.}$$

3 Equilibrium of a Particle

Coplanar Force Systems

3-1. The particle is subjected to three forces. Determine the magnitudes of F_1 and F_2.

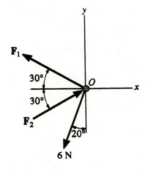

Solution

The figure represents the free-body diagram. Apply the equations of equilibrium.

$$\overset{+}{\underset{\rightarrow}{}} \Sigma F_x = 0; \underline{\hspace{8cm}}$$

$$+\uparrow \Sigma F_y = 0; \underline{\hspace{8cm}}$$

Thus,

$$F_1 = 4.45 \text{ N}, \quad F_2 = 6.82 \text{ N} \qquad\qquad \textbf{Ans.}$$

3 - 2. Determine the magnitude and direction θ of the force **F** that acts along link AB. The suspended mass is 10 kg. Neglect the size of the pulley at A.

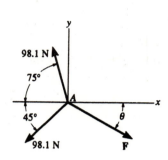

Solution

The free - body diagram is drawn first. Apply the equations of equilibrium.

$$\overset{+}{\rightarrow} \Sigma F_x = 0;\underline{\hspace{9cm}}$$

$$+ \uparrow \Sigma F_y = 0;\underline{\hspace{9cm}}$$

Thus,

$$F_{AB} = 98.1 \text{ N} \qquad \qquad \textbf{Ans.}$$

$$\theta = 15.0° \qquad \qquad \textbf{Ans.}$$

3 - 3. A block is suspended from a single elastic cord ABC, which has a stiffness of 5 N/m and an unstretched length of 4 m. If $d = 1.5$ m and the ring at B is free to slide over the cord, show that cord length AB must be equal to CB and determine the mass of the block.

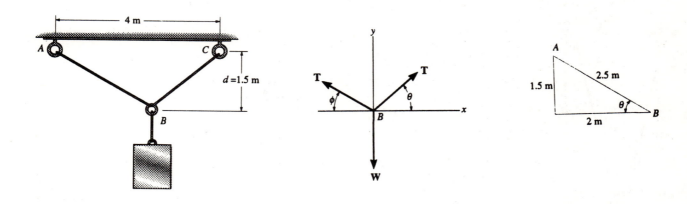

Solution

The free - body diagram is drawn first. Apply the equations of equilibrium.

$$\stackrel{+}{\rightarrow} \Sigma F_x = 0; \underline{\hspace{7cm}} \quad ; \theta = \phi$$

$$+ \uparrow \Sigma F_y = 0; \underline{\hspace{6cm}}$$

For $d = 1.5$ m, the length AB becomes 2.5 m as shown in the sketch. Therefore the cord is stretched $2(2.5 \text{ m}) - 4 \text{ m} = 1 \text{ m}$

$$F = kx; \quad T = \underline{\hspace{5cm}} = 5.00 \text{ N}$$

Since

$$\theta = \underline{\hspace{4cm}} = 36.87°$$

Then

$$W = 6.00 \text{ N}$$

$$m = \frac{6.00}{9.81} = 0.612 \text{ kg} \qquad \qquad \textbf{Ans.}$$

37

3 - 4. A "scale" is constructed with a 4 - m - long cord and the 10 - kg block D. The cord is fixed to a pin at A and passes over two *small* pulleys at B and C. Determine the mass of the suspended block E if the system is in equilibrium when $d = 1.5$ m.

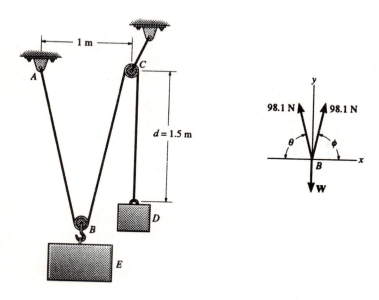

Solution

The free - body diagram is drawn first. Apply the equations of equilibrium.

$$\xrightarrow{+} \Sigma F_x = 0; \underline{\hspace{6cm}}$$

$$\theta = \phi$$

$$+\uparrow \Sigma F_y = 0; \underline{\hspace{6cm}}$$

The angle θ is determined from the geometry of the triangle.

$$W = 196.2 \sin\theta = 196.2(\frac{\sqrt{(1.25)^2 - (0.5)^2}}{1.25})$$

$$W = 179.82 \text{ N}$$

$$m = \frac{179.82}{9.81} = 18.3 \text{ kg} \qquad \qquad \textbf{Ans.}$$

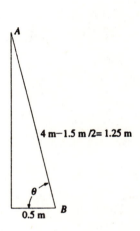

4 m−1.5 m /2= 1.25 m

0.5 m

3 - 5. Determine the force in each cable and the force **F** needed to hold the 4 - kg block in the position shown.

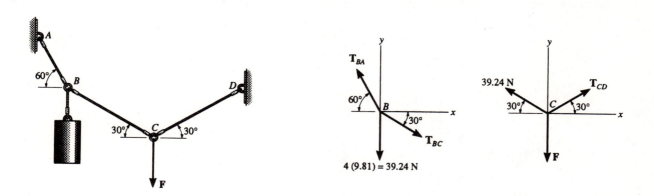

Solution

The free - body diagram of point B is used to determine the forces in BA and BC.
Apply the equations of equilibrium.

$\xrightarrow{+} \Sigma F_x = 0;$ _____

$+\uparrow \Sigma F_y = 0;$ _____

Thus,

$T_{BC} = 39.24 \text{ N} = 39.2 \text{ N}$ **Ans.**

$T_{BA} = 68.0 \text{ N}$ **Ans.**

Using the result for T_{BC}, the free - body diagram of point C can now be used to determine the forces T_{CD} and F.

$\xrightarrow{+} \Sigma F_x = 0;$ _____

$+\uparrow \Sigma F_y = 0;$ _____

Thus,

$T_{CD} = 32.9 \text{ N}, \quad F = 39.2 \text{ N}$ **Ans.**

3 - 6. A continuous cord of total length 4 m is wrapped around the *small* frictionless pulleys at *A*, *B*, *C*, and *D*. If the stiffness of each spring is $k = 500$ N/m and each spring is streteched 0.3 m, determine the mass *m* of each block. Neglect the weight of the pulleys and cords. The springs are unstretched when $d = 2$ m.

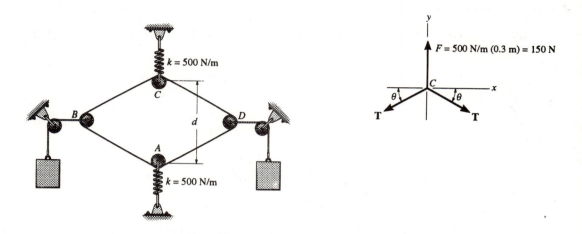

Solution

The free - body diagram of point *C* is considered to relate the known spring force to the force in the continuous cord.

$$+\uparrow \Sigma F_y = 0;$$ _____

$$T = \frac{75}{\sin\theta} \qquad (1)$$

As stated, when $d = 2$ m the springs are unstretched and the tension in the cord is zero. When the springs are stretched 0.3 m, then

$$d = (2 - 2(0.3)) = 1.4 \text{ m}$$

From the geometry of the triangle,

$$\theta = \sin^{-1}\frac{0.7}{1} = 44.3°$$

From Eq. (1),

$$T = 107.1 \text{ N}$$

Using the free - body diagram of point *B*,

$$\xrightarrow{+} \Sigma F_x = 0;$$ _____

Thus,

$$m = 15.6 \text{ kg} \qquad \qquad \textbf{Ans.}$$

Three - Dimensional Force Systems

3 - 7. The particle is subjected to four forces. Force **F** lies in the $x - y$ plane and **P** lies in the $y - z$ plane. Determine the magnitudes of **F**, **R**, and **P** required for equilibrium.

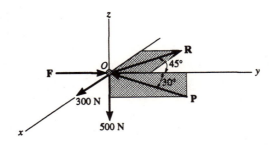

Solution

The figure represents the free - body diagram . The scalar components of each force can be determined directly from the diagram. Apply the equations of equilibrium.

$\Sigma F_x = 0;$ _____

$\Sigma F_y = 0;$ _____

$\Sigma F_z = 0;$ _____

Thus,

$P = 1 \text{ kN}, \quad R = 424 \text{ N}, \quad F = 566 \text{ N}$ **Ans.**

3 - 8. Determine the magnitudes of the forces **P**, **R**, and **F** required for equilibrium of the particle.

Solution

The figure represents the free-body diagram of the particle. Express each force as a Cartesian vector.

$R = $  _____

$F_1 = $ _____

$F_2 = $ _____

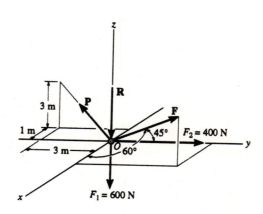

$$P = P[(\frac{}{\sqrt{19}})i + (\frac{}{\sqrt{19}})j + (\frac{}{\sqrt{19}})k]$$

$$= -0.2294Pi - 0.6882Pj + 0.6882Pk$$

Write the relationship used to determine γ for F.

_____

As shown in the figure, $\gamma < 90°$

$$\gamma = 60°$$

$$F = F(\underline{\quad\quad} i + \underline{\quad\quad} j + \underline{\quad\quad} k)$$

$$= 0.5Fi + 0.7071Fj + 0.5k$$

Equilibrium

$$\Sigma F_x = 0; \quad -0.2294P + 0.5F = 0$$
$$\Sigma F_y = 0; \quad 400 - 0.6882P + 0.7071F = 0$$
$$\Sigma F_z = 0; \quad -R - 600 + 0.6882P + 0.5F = 0$$

Thus,

$$P = 1.10 \text{ kN}, \quad F = 504 \text{ N}, \quad R = 409 \text{ N} \qquad \textbf{Ans.}$$

3 - 9. The ends of the three cables are attached to a ring at A and to the edge of a uniform 150 - kg plate. Determine the tension in each of the cables for equilibrium.

Solution

The free - body diagram of point A is drawn first, then each force is expressed as a Cartesian vector.

$$\mathbf{F_B} = F_B(\frac{\quad}{14} \mathbf{i} + \frac{\quad}{14} \mathbf{j} + \frac{\quad}{14})$$

$$\mathbf{F_C} = F_C(\frac{\quad}{14} \mathbf{i} + \frac{\quad}{14} \mathbf{j} + \frac{\quad}{14} \mathbf{k})$$

$$\mathbf{F_D} = F_D(\frac{\quad}{14} \mathbf{i} + \frac{\quad}{14} \mathbf{j} + \frac{\quad}{14} \mathbf{k})$$

$$\mathbf{P} = \{1471.5\mathbf{k}\} \text{ N}$$

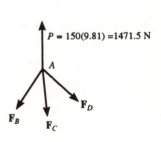

Equilibrium

$$\Sigma F_x = 0; \quad \frac{4}{14}F_B - \frac{6}{14}F_C - \frac{4}{14}F_D = 0$$

$$\Sigma F_y = 0; \quad -\frac{6}{14}F_B - \frac{4}{14}F_C + \frac{6}{14}F_D = 0$$

$$\Sigma F_z = 0; \quad -\frac{12}{14}F_B - \frac{12}{14}F_C - \frac{12}{14}F_D + 1471.5 = 0$$

Thus,

$$F_B = 858 \text{ N}, \quad F_C = 0, \quad F_D = 858 \text{ N}$$

Ans.

43

3-10. Determine the force in the cable required to support the vertical load of $F = 500$ N. The elastic cord OA has unstretched length of 0.2 m and a stiffness of $k_{OA} = 350$ N/m.

Solution

The free-body diagram of point O is drawn first. Each force is expressed as a Cartesian vector.

$$F_A = kx = 350(1 - 0.2) = 280 \text{ N}$$

$$\mathbf{F}_A = \{-280\mathbf{j}\} \text{ N}$$

$$\mathbf{F} = \{-500\mathbf{k}\} \text{ N}$$

$$\mathbf{F}_B = F_B(\frac{-2}{6}\mathbf{i} - \frac{4}{6}\mathbf{j} + \frac{4}{6}\mathbf{k})$$

$$= -0.333F_B\mathbf{i} - 0.667F_B\mathbf{j} + 0.667F_B\mathbf{k}$$

$$\mathbf{F}_C = F_C(\frac{-4}{5}\mathbf{i} + \frac{3}{5}\mathbf{k})$$

$$= -0.8F_C\mathbf{i} + 0.6F_C\mathbf{k}$$

$$\mathbf{F}_D = F_D(\frac{2}{6}\mathbf{i} + \frac{4}{6}\mathbf{j} + \frac{4}{6}\mathbf{k})$$

$$= 0.333F_D\mathbf{i} + 0.667F_D\mathbf{j} + 0.667F_D\mathbf{k}$$

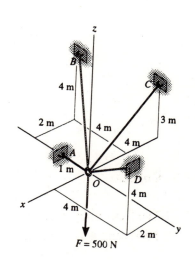

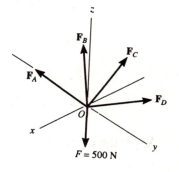

Apply the equations of equilibrium.

$\Sigma F_x = 0;$ _____

$\Sigma F_y = 0;$ _____

$\Sigma F_z = 0;$ _____

Thus,

$$F_B = 86.25 \text{ N}, \quad F_C = 175 \text{ N}, \quad F_D = 506 \text{ N} \qquad \textbf{Ans.}$$

44

3 - 11. Determine the force acting along the axis of each strut necessary to hold the 20 - kg cylinder in equilibrium.

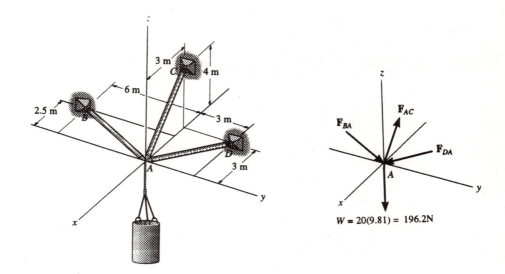

Solution

The free - body diagram of point A is first drawn. The sense of direction of the unknown forces has been assumed. Expressing each force as a Cartesian vector,

$$\mathbf{F}_{BA} = F_{BA} \left(\frac{2.5}{6.5} \mathbf{i} + \frac{6}{6.5} \mathbf{j} \right) = 0.385 F_{BA} \mathbf{i} + 0.923 F_{BA} \mathbf{j}$$

$$\mathbf{F}_{DA} = F_{DA} \left(\frac{3}{4.243} \mathbf{i} - \frac{3}{4.243} \mathbf{j} \right) = 0.707 F_{DA} \mathbf{i} - 0.707 F_{DA} \mathbf{j}$$

$$\mathbf{F}_{AC} = F_{AC} \left(\frac{-3}{5} \mathbf{i} + \frac{4}{5} \mathbf{k} \right) = -0.6 F_{AC} \mathbf{i} + 0.8 F_{AC} \mathbf{k}$$

$$\mathbf{W} = -196.2 \, \mathbf{k}$$

Apply the equations of equilibrium,

$$\Sigma F_x = 0; \underline{\hspace{8cm}}$$

$$\Sigma F_y = 0; \underline{\hspace{8cm}}$$

$$\Sigma F_z = 0; \underline{\hspace{8cm}}$$

Thus,

$$F_{BA} = 113 \, \text{N}, \quad F_{AC} = 245 \, \text{N}, \quad F_{DA} = 147 \, \text{N} \qquad \qquad \textbf{Ans.}$$

4 Force System Resultants

Moment of a Force

4 - 1. In each case, determine the moment of the force at A about point P.

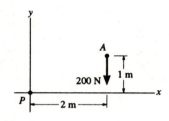

(a)

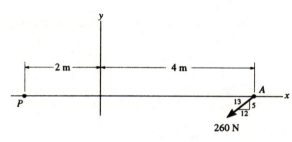

(b)

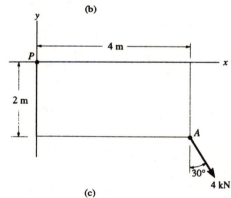

(c)

Solution

a) $\left(+ M_P = \rule{5cm}{0.4pt} = 400 \text{ N} \cdot \text{m}$ **Ans.**

b) $\left(+ M_P = \rule{5cm}{0.4pt} = 600 \text{ N} \cdot \text{m}$ **Ans.**

c) $\left(+ M_P = \rule{5cm}{0.4pt} = 9.86 \text{ kN} \cdot \text{m}$ **Ans.**

4 - 2. Determine the moment of each of the three forces about point A on the beam.

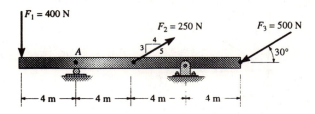

Solution

$M_1 = $ _____ $= 1600 \text{ N} \cdot \text{m}$　　　　**Ans.**

$M_2 = $ _____ $= 600 \text{ N} \cdot \text{m}$　　　　**Ans.**

$M_3 = $ _____ $= 3000 \text{ N} \cdot \text{m}$　　　　**Ans.**

4 - 3. In each case, determine the moment of the force at A about point P.

Solution

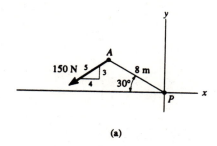

(a)

$\zeta + M_P = $ _____

$M_P = 1.10 \text{ kN} \cdot \text{m}$ **Ans.**

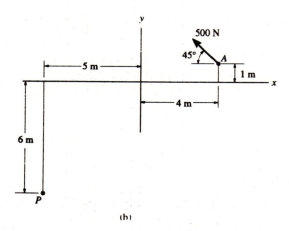

(b)

$\zeta + M_P = $ _____

$M_P = 5.66 \text{ kN} \cdot \text{m}$ **Ans.**

48

4 - 4. Determine the direction θ ($0° \le \theta \le 180°$) of the 30 - N force so that it produces (a) the maximum moment about point A and (b) the minimum moment about point A. Compute the moment in each case.

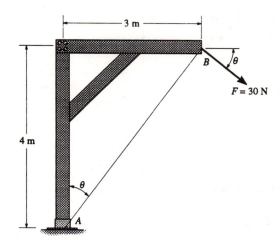

Solution

a) For maximum moment set $\mathbf{F} \perp$ to a line from A to B.

$$\theta = \tan^{-1} (\text{————}) = 36.9°$$ **Ans.**

$$(M_A)_{max} = \underline{\hspace{3cm}} = 150 \text{ N} \cdot \text{m}$$ **Ans.**

b) Minimum moment occurs when $\mathbf{F}$ passes through A.

$$\theta = 90° + \tan^{-1}\frac{3}{4} = 127°$$ **Ans.**

$$(M_A)_{min} = 0$$ **Ans.**

49

4 - 5. Determine the moment of the force at A about point P. Use a vector analysis and express the result in Cartesian vector form.

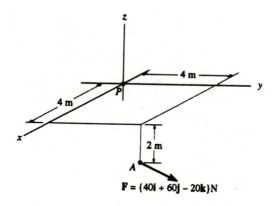

z

4 m

y

4 m

P

x

2 m

A

$F = \{40i + 60j - 20k\}N$

Solution

$$M_P = \begin{vmatrix} i & j & k \\ \underline{} & \underline{} & \underline{} \\ \underline{} & \underline{} & \underline{} \end{vmatrix}$$

$M_P = \{40i + 80k\} \, N \cdot m$ **Ans.**

4 - 6. Determine the moment of the force at *A* about point *P*. Use a vector analysis and express the result in Cartesian vector form.

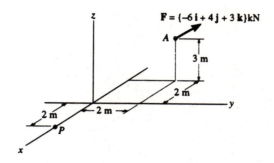

Solution

$$M_P = \begin{vmatrix} \mathbf{i} & \mathbf{j} & \mathbf{k} \\ \rule{1em}{0.4pt} & \rule{1em}{0.4pt} & \rule{1em}{0.4pt} \\ \rule{1em}{0.4pt} & \rule{1em}{0.4pt} & \rule{1em}{0.4pt} \end{vmatrix}$$

$$M_P = \{-6\mathbf{i} - 6\mathbf{j} - 4\mathbf{k}\} \text{ kN} \cdot \text{m} \qquad\qquad \textbf{Ans.}$$

4 - 7. Determine the moment of the force at A about point P. Use a vector analysis and express the result in Cartesian vector form.

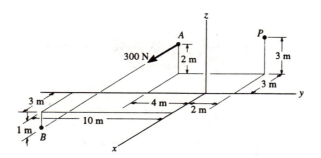

Solution

$$F = 300 \text{ N}(\frac{6}{9}\mathbf{i} - \frac{6}{9}\mathbf{j} - \frac{3}{9}\mathbf{k}) = \{200\mathbf{i} - 200\mathbf{j} - 100\mathbf{k}\} \text{ N}$$

Show two possible position vectors that can be used.

$$M_P = \begin{vmatrix} \mathbf{i} & \mathbf{j} & \mathbf{k} \\ \underline{\quad} & \underline{\quad} & \underline{\quad} \\ \underline{\quad} & \underline{\quad} & \underline{\quad} \end{vmatrix} = \begin{vmatrix} \mathbf{i} & \mathbf{j} & \mathbf{k} \\ \underline{\quad} & \underline{\quad} & \underline{\quad} \\ \underline{\quad} & \underline{\quad} & \underline{\quad} \end{vmatrix}$$

$$M_P = \{400\mathbf{i} - 200\mathbf{j} + 1200\mathbf{k}\} \text{ N} \cdot \text{m} \qquad \textbf{Ans.}$$

4 - 8. The force of 140 N acts on the end of a beam and is directed towards point B as shown. Determine the moment of this force at point A.

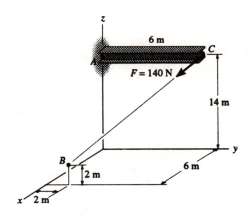

Solution

$$\mathbf{r}_{AC} = \underline{\hspace{10cm}}$$

$$\mathbf{F} = 140 \text{ N}(\frac{}{14})\mathbf{i} + (\frac{}{14})\mathbf{j} + (\frac{}{14})\mathbf{k}$$
$$= \{60\mathbf{i} - 40\mathbf{j} - 120\mathbf{k}\} \text{ N}$$

$$\mathbf{M}_A = \begin{vmatrix} \mathbf{i} & \mathbf{j} & \mathbf{k} \\ \underline{\quad} & \underline{\quad} & \underline{\quad} \\ \underline{\quad} & \underline{\quad} & \underline{\quad} \end{vmatrix}$$

$$\mathbf{M}_A = \{-720\mathbf{i} - 360\mathbf{k}\} \text{ N} \cdot \text{m} \qquad\qquad \textbf{Ans.}$$

4 - 9. The pole supports a vertical force of $F = 22$ N. Using Cartesian vectors, determine the moment of the force about the base of the pole at A.

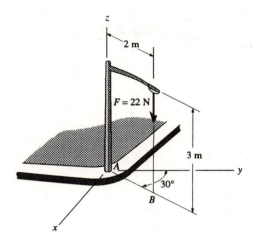

Solution

$$\mathbf{r}_{AB} = \underline{\hspace{5cm}}$$

$$= \{1.0\mathbf{i} + 1.732\mathbf{j}\} \text{ m}$$

$$\mathbf{F} = \underline{\hspace{6cm}}$$

$$\mathbf{M}_A = \begin{vmatrix} \underline{\quad} & \underline{\quad} & \underline{\quad} \\ \underline{\quad} & \underline{\quad} & \underline{\quad} \\ \underline{\quad} & \underline{\quad} & \underline{\quad} \end{vmatrix}$$

$$\mathbf{M}_A = \{-38.1\mathbf{i} + 22\mathbf{j}\} \text{ N} \cdot \text{m} \qquad \textbf{Ans.}$$

Moment of a Force About a Specified Axis

4 - 10. Determine the projection of the moment caused by the force about the *aa* axis.

Solution

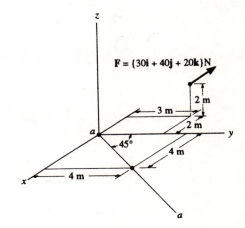

$$F = \{30i + 40j + 20k\} \text{ N}$$

List two possible unit vectors that can be used.

$$u = \rule{6cm}{0.4pt}$$

$$u = \rule{6cm}{0.4pt}$$

List two posssible position vectors that can be used.

$$r = \rule{6cm}{0.4pt}$$

$$r = \rule{6cm}{0.4pt}$$

Selecting from the above, enter one possible solution.

$$M_P = \begin{vmatrix} \rule{0.6cm}{0.4pt} & \rule{0.6cm}{0.4pt} & \rule{0.6cm}{0.4pt} \\ \rule{0.6cm}{0.4pt} & \rule{0.6cm}{0.4pt} & \rule{0.6cm}{0.4pt} \\ \rule{0.6cm}{0.4pt} & \rule{0.6cm}{0.4pt} & \rule{0.6cm}{0.4pt} \end{vmatrix} = |56.6 \text{ kN} \cdot \text{m}| \qquad \textbf{Ans.}$$

4 - 11. Determine the projection of the moment caused by the force about the *aa* axis.

Solution

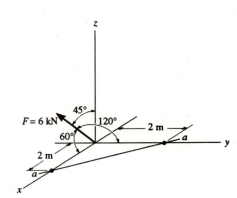

$$F = \underline{\hspace{6cm}}$$

$$= \{3\mathbf{i} - 3\mathbf{j} + 4.243\mathbf{k}\} \text{ kN}$$

List two possible unit vectors that can be used.

$$\mathbf{u} = \underline{\hspace{6cm}}$$

$$\mathbf{u} = \underline{\hspace{6cm}}$$

List two posssible position vectors that can be used.

$$\mathbf{r} = \underline{\hspace{6cm}}$$

$$\mathbf{r} = \underline{\hspace{6cm}}$$

Selecting from the above, enter one possible solution.

$$M_P = \begin{vmatrix} \underline{\quad} & \underline{\quad} & \underline{\quad} \\ \underline{\quad} & \underline{\quad} & \underline{\quad} \\ \underline{\quad} & \underline{\quad} & \underline{\quad} \end{vmatrix} = |6 \text{ kN} \cdot \text{m}| \qquad \textbf{Ans.}$$

4 - 12. A force of 50 N is applied to the plate as shown. Determine the projection of the moment of this force about the z axis. Use a scalar analysis.

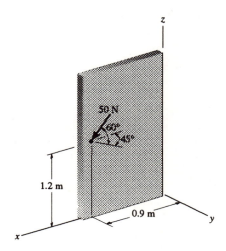

Solution

Force component perpendicular to face of plate is

$$F_y = \text{_____}$$

$$= 17.68 \text{ N}$$

Thus, if $\mathbf{u} = +\mathbf{k}$, then

$$M_z = \text{_____}$$

$$= -15.9 \text{ N} \cdot \text{m} \qquad\qquad\qquad\qquad \textbf{Ans.}$$

The negative sign indicates that $\mathbf{M}_z$ acts downward.

Moment of a Couple

4 - 13. Determine the magnitude and direction of the couple moment.

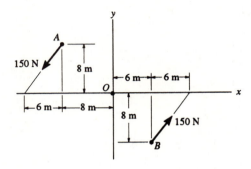

Solution

Sum moments about O,

$$M_c = \underline{\hspace{5cm}}$$

$$= 3.12 \text{ kN} \cdot \text{m} \qquad\qquad \textbf{Ans.}$$

Sum moments about A,

$$M_c = \underline{\hspace{5cm}}$$

$$= 3.12 \text{ kN} \cdot \text{m} \qquad\qquad \textbf{Ans.}$$

Sum moments about B,

$$M_c = \underline{\hspace{5cm}}$$

$$= 3.12 \text{ kN} \cdot \text{m} \qquad\qquad \textbf{Ans.}$$

4 - 14. Determine the magnitude and direction of the couple moment.

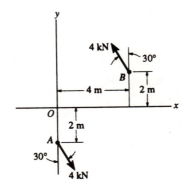

Solution

Sum moments about O,

$$M_c = \underline{\hspace{6cm}}$$

$$= 21.9 \text{ kN} \cdot \text{m} \qquad\qquad \textbf{Ans.}$$

Sum moments about A,

$$M_c = \underline{\hspace{6cm}}$$

$$= 21.9 \text{ kN} \cdot \text{m} \qquad\qquad \textbf{Ans.}$$

Sum moments about B,

$$M_c = \underline{\hspace{6cm}}$$

$$= 21.9 \text{ kN} \cdot \text{m} \qquad\qquad \textbf{Ans.}$$

4 - 15. A twist of 4 N · m is applied to the handle of the screwdriver. Resolve this couple moment into a pair of couple forces **F** exerted on the handle and **P** exerted on the blade.

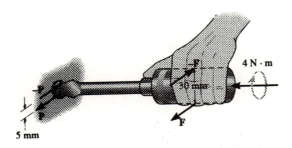

Solution

$F = $ _____

$\qquad F = 133$ N   **Ans.**

$P = $ _____

$\qquad P = 800$ N   **Ans.**

4-16. The main beam along the wing of an airplane is swept back at an angle of 25°. From load calculations it is determined that the beam is subjected to couple moments $M_x = 25$ kN · m and $M_y = 17$ kN · m. Determine the equivalent couple moments created about the x' and y' axes.

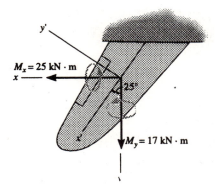

Solution

Resolve each vector into x' and y' components and add the components.

$M_x' = $ _____

$M_x' = 26.0$ kN · m   **Ans.**

$M_y' = $ _____

$M_y' = 15.5$ kN · m   **Ans.**

4 - 17. Determine the couple moment. Use a vector analysis and express the result as a Cartesian vector.

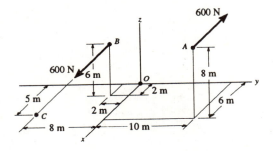

Solution

Summing moments of the forces about point A we have

$$\mathbf{r}_{AB} = \underline{\hspace{6cm}}$$

$$\mathbf{F} = 600 \text{ N}(\frac{-}{9}\mathbf{i} + \frac{-}{9}\mathbf{j} + \frac{-}{9}\mathbf{k})$$

$$= \{200\mathbf{i} - 400\mathbf{j} - 400\mathbf{k}\} \text{ N}$$

$$\mathbf{M}_C = \mathbf{r}_{AB} \times \mathbf{F} = \begin{vmatrix} \mathbf{i} & \mathbf{j} & \mathbf{k} \\ \underline{\hspace{0.5cm}} & \underline{\hspace{0.5cm}} & \underline{\hspace{0.5cm}} \\ \underline{\hspace{0.5cm}} & \underline{\hspace{0.5cm}} & \underline{\hspace{0.5cm}} \end{vmatrix}$$

$$\mathbf{M}_C = \{4000\mathbf{i} - 2000\mathbf{j} + 4000\mathbf{k}\} \text{ N} \cdot \text{m}$$

Note that this result can also be determined by summing moments about point B (one force), or about point O (both forces).

4 - 18. Express the moment of the couple acting on the pipe in Cartesian vector form.

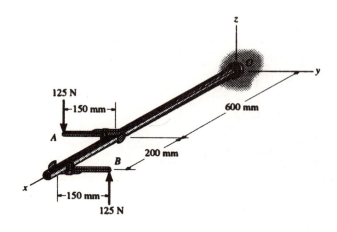

Solution

Summing the moments about point A, we have

$$\mathbf{r}_{AB} = \underline{\hspace{6cm}}$$

$$\mathbf{F}_B = \underline{\hspace{6cm}}$$

$$\mathbf{M}_c = \begin{vmatrix} \mathbf{i} & \mathbf{j} & \mathbf{k} \\ \underline{\hspace{1cm}} & \underline{\hspace{1cm}} & \underline{\hspace{1cm}} \\ \underline{\hspace{1cm}} & \underline{\hspace{1cm}} & \underline{\hspace{1cm}} \end{vmatrix}$$

$$\mathbf{M}_c = \{37.5\mathbf{i} - 25\mathbf{j}\} \text{ N} \cdot \text{m} \qquad\qquad \textbf{Ans.}$$

This same result can also be determined from summing the moment about point A (one force) or about point O (two forces).

Simplification of a Force System

4 - 19. Replace the force at A by an equivalent force and couple moment at P.

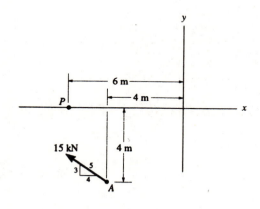

Solution

$$F = \underline{\hspace{4cm}} \qquad\qquad \textbf{Ans.}$$

$$\curvearrowright + M_p = \underline{\hspace{5cm}}$$

$$M_p = 30.0 \text{ kN} \cdot \text{m} \qquad\qquad \textbf{Ans.}$$

4 - 20. The structural connection is subjected to the 8 kN force. Replace this force by an equivalent force and couple moment acting at the center of the bolt group, *O*.

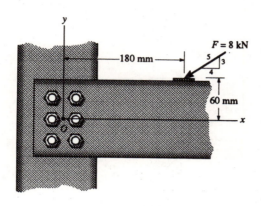

Solution

$F =$ _____ **Ans.**

$M_O =$ _____

$M_O = 480\ \text{N} \cdot \text{m}$ **Ans.**

4 - 21. Replace the force and couple system by an equivalent resultant force and couple moment acting at point P.

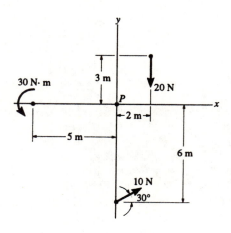

Solution

$$\overset{+}{\rightarrow} F_{R_x} = \Sigma F_x = \underline{\hspace{6cm}} = 8.66 \text{ N} \rightarrow$$

$$+ \uparrow F_{R_y} = \Sigma F_y = \underline{\hspace{6cm}} = -15 \text{ N} = 15 \text{ N} \downarrow$$

$$\zeta + M_{RP} = \Sigma M_P = \underline{\hspace{7cm}}$$

$$M_{RP} = 42.0 \text{ N} \cdot \text{m} \qquad\qquad \textbf{Ans.}$$

$$F_R = \sqrt{(8.66)^2 + (-15)^2} = 17.3 \text{ N} \qquad\qquad \textbf{Ans.}$$

$$\theta = \tan^{-1}(\frac{15}{8.66}) = 60° \qquad\qquad \textbf{Ans.}$$

4 - 22. Replace the force and couple system by an equivalent resultant force and couple moment acting at point P.

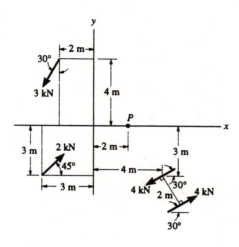

Solution

$$\overset{+}{\rightarrow} F_{R_x} = \Sigma F_x = \underline{\hspace{5cm}} = -0.0858 \text{ kN} = 0.0858 \text{ kN} \leftarrow$$

$$+\uparrow F_{R_y} = \Sigma F_y = \underline{\hspace{5cm}} = -1.18 \text{ kN} = 1.18 \text{ kN} \downarrow$$

$$\zeta + M_{R\,P} = \Sigma M_P = \underline{\hspace{6cm}}$$

$$M_{RP} = 21.6 \text{ kN} \cdot \text{m} \qquad\qquad \textbf{Ans.}$$

$$F_R = \sqrt{(-0.0858)^2 + (-1.18)^2} = 1.19 \text{ N} \qquad\qquad \textbf{Ans.}$$

$$\theta = \tan^{-1}\left(\frac{1.18}{0.0858}\right) = 85.9° \qquad\qquad \textbf{Ans.}$$

4 - 23. Replace the loading system by an equivalent resultant force and couple moment at point O.

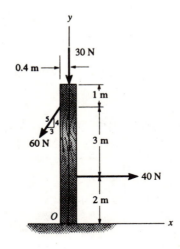

Solution

$$\overset{+}{\rightarrow} F_{R_x} = \Sigma F_x ; \ F_{R_x} = \underline{\hspace{5cm}} = 4 \ N \rightarrow$$

$$+ \uparrow F_{R_y} = \Sigma F_y ; \ F_{R_y} = \underline{\hspace{5cm}} = -78 \ N = 78 \ N \downarrow$$

$$F_R = \sqrt{(4)^2 + (-78)^2} = 78.1 \ N \qquad\qquad \textbf{Ans.}$$

$$\theta = \tan^{-1} \frac{78}{4} = 87.1° \qquad\qquad \textbf{Ans.}$$

$$\zeta + M_{RO} = \Sigma M_O ; \ M_{R_O} = \underline{\hspace{5cm}}$$

$$M_{R\,O} = 119 \ N \cdot m \qquad\qquad \textbf{Ans.}$$

4 - 24. Replace the force at A by an equivalent resultant force and couple moment at P. Express the results in Cartesian vector form.

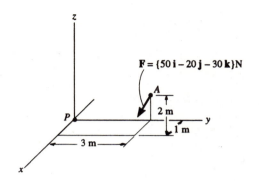

Solution

$\mathbf{F}_R = $ _____   **Ans.**

$$\mathbf{M}_{RP} = \begin{vmatrix} \mathbf{i} & \mathbf{j} & \mathbf{k} \\ — & — & — \\ — & — & — \end{vmatrix}$$

$\mathbf{M}_{RP} = \{-50\mathbf{i} + 130\mathbf{j} - 170\mathbf{k}\} \ \mathrm{N \cdot m}$ **Ans.**

4-25. Replace the force at A by an equivalent resultant force and couple moment at P. Express the results in Cartesian vector form.

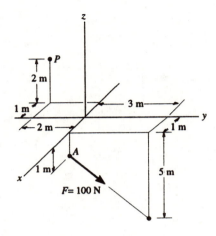

Solution

$$\mathbf{F}_R = 100 \text{ N}(\frac{}{5}\mathbf{i} + \frac{}{5}\mathbf{j} + \frac{}{5}\mathbf{k})$$

$$= \{60\mathbf{j} - 80\mathbf{k}\} \text{ N} \qquad\qquad \textbf{Ans.}$$

$$\mathbf{M}_{RP} = \begin{vmatrix} \mathbf{i} & \mathbf{j} & \mathbf{k} \\ \text{---} & \text{---} & \text{---} \\ \text{---} & \text{---} & \text{---} \end{vmatrix}$$

$$\mathbf{M}_{RP} = \{20\mathbf{i} + 160\mathbf{j} + 120\mathbf{k}\} \text{ N} \cdot \text{m} \qquad\qquad \textbf{Ans.}$$

4 - 26. The resultant force of a wind loading acts perpendicular to the face of the sign as shown. Replace this force by an equivalent resultant force and couple moment acting at point O.

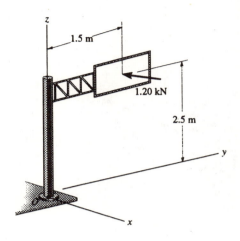

Solution

$$\mathbf{F}_R = \text{_____} \qquad \textbf{Ans.}$$

$$M_{RO} = \begin{vmatrix} \mathbf{i} & \mathbf{j} & \mathbf{k} \\ \text{\_\_\_} & \text{\_\_\_} & \text{\_\_\_} \\ \text{\_\_\_} & \text{\_\_\_} & \text{\_\_\_} \end{vmatrix}$$

$$M_{RO} = \{-3\mathbf{j} + 1.8\mathbf{k}\} \text{ kN} \cdot \text{m} \qquad \textbf{Ans.}$$

4 - 27. Replace the force by an equivalent resultant force and couple moment at point A.

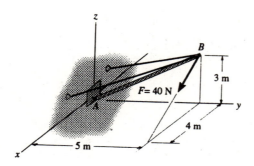

Solution

$$F = (\underline{\hspace{2cm}})[(\frac{}{5})i + (\frac{}{5})j + (\frac{}{5})k]$$

$$= \{32i - 24k\} \text{ N} \qquad\qquad\qquad \textbf{Ans.}$$

$$M_{RA} = r_{AB} \times F = \begin{vmatrix} i & j & k \\ \text{---} & \text{---} & \text{---} \\ \\ \text{---} & \text{---} & \text{---} \end{vmatrix}$$

$$M_{RA} = \{-120i + 96j - 160k\} \text{ N} \cdot \text{m} \qquad\qquad \textbf{Ans.}$$

4 - 28. Replace the two forces acting on the tree branches by an equivalent resultant force and couple moment acting at point O.

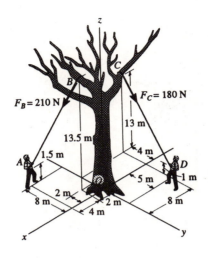

Solution

$$\mathbf{F}_B = 210 \text{ N}[(\frac{}{14})\mathbf{i} + (\frac{}{14})\mathbf{j} + (\frac{}{14})\mathbf{k} = \{60\mathbf{i} - 90\mathbf{j} - 180\mathbf{k}\} \text{ N}$$

$$\mathbf{F}_C = 180 \text{ N}[(\frac{}{15})\mathbf{i} + (\frac{}{15})\mathbf{j} + (\frac{}{15})\mathbf{k}] = \{108\mathbf{j} - 144\mathbf{k}\} \text{ N}$$

$\mathbf{F}_R = $ _____   **Ans.**

$\mathbf{M}_{RO} = $ _____ $\times (60\mathbf{i} - 90\mathbf{j} - 180\mathbf{k})$

$+$ _____ $\times (108\mathbf{j} - 144\mathbf{k})$

$\mathbf{M}_{RO} = \{747\mathbf{i} + 18\mathbf{j} - 924\mathbf{k}\} \text{ N} \cdot \text{m}$ **Ans.**

Reduction of a Simple Distributed Loading

4 - 29. Replace this system of three forces by a single resultant force and specify its vertical location from point O.

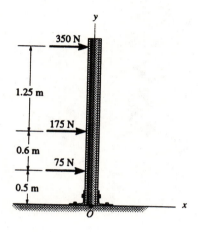

Solution

$$\stackrel{+}{\to} F_{R_x} = \Sigma F_x \; ; \; F_{R_x} = \text{\underline{\hspace{6cm}}} = 600 \text{ N} \quad \textbf{Ans.}$$

$$\zeta + M_{R_O} = \Sigma M_O \; ; \; -600(d) = \text{\underline{\hspace{5cm}}}$$

$$d = 1.75 \text{ m} \qquad\qquad \textbf{Ans.}$$

4 - 30. Determine the location (x, y) of the 4 - kN force so that all four forces create a single resultant force acting through the plate's center O.

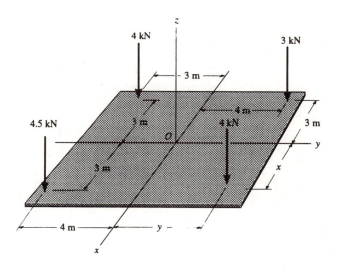

Solution

$M_{R_y} = \Sigma M_y ;$ _____

$x = 1.875$ m   **Ans.**

$M_{R_x} = \Sigma M_x ;$ _____

$y = 4.50$ m   **Ans.**

4 - 31. A force and couple act on the pipe assembly. Replace this system by an equivalent single resultant force. Specify the location of the resultant force along the y axis, measured from A. The assembly lies in the $x - y$ plane.

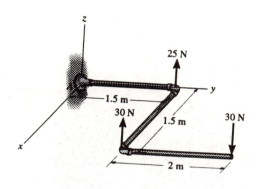

Solution

$$+\uparrow F_{R_z} = \Sigma F_z; \quad F_{R_z} = \underline{\hspace{6cm}} \quad \textbf{Ans.}$$

$$M_{A_x} = \Sigma M_x; \underline{\hspace{7cm}}$$

$$y = -0.9 \text{ m} \qquad \textbf{Ans.}$$

The result indicates that the resultant force *does not* act on the assembly.

4 - 32. Three parallel forces act on the rim of the tube. If it is required that the resultant force passes through the central axis O, determine the magnitude of $\mathbf{F}_C$ and its location θ on the rim. What is the magnitude of th resultant force?

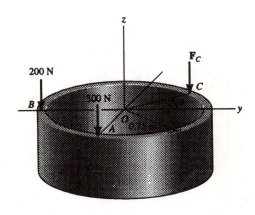

Solution

$$+\uparrow F_z = \Sigma F_z; \qquad F_R = -200 - 300 - F_C$$

Since F_R is to be located at point O, then

$$M_{R_x} = \Sigma M_x;$$

$$0 = \underline{\hspace{10cm}}$$

$$M_{R_y} = \Sigma M_y;$$

$$0 = \underline{\hspace{10cm}}$$

Thus,

$$F_C = 361 \text{ N}, \quad \theta = 56.3°, \quad F_R = -861 \text{ N} = 861 \text{ N} \downarrow \qquad\qquad \textbf{Ans.}$$

4 - 33. Replace the loading by an equivalent resultant force and couple moment acting at point O.

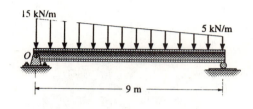

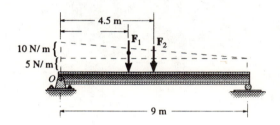

Solution

$$F_1 = \rule{7cm}{0.4pt} = 45 \text{ kN}$$

$$F_2 = \rule{7cm}{0.4pt} = 45 \text{ kN}$$

$$+ \uparrow F_{R_y} = \Sigma F_y ; \quad F_{R_y} = \rule{5cm}{0.4pt}$$

$$= -90 \text{ kN} = 90 \text{ kN} \downarrow \qquad \qquad \textbf{Ans.}$$

$$\zeta + M_{R_O} = \Sigma M_O ; \quad M_{R_O} = \rule{5cm}{0.4pt}$$

$$= -337.5 \text{ kN} \cdot \text{m} = 337.5 \text{ kN} \cdot \text{m} \qquad \qquad \textbf{Ans.}$$

4 - 34. Simplify the distributed loading to a single resultant force and specify the magnitude and location of the force measured from A.

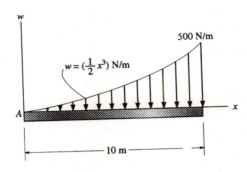

Solution

$$+ \downarrow F_R = \int \underline{\hspace{8cm}}$$

$$F_R = 1250 \text{ N} \downarrow \qquad\qquad \textbf{Ans.}$$

$$\curvearrowleft + M_{R_A} = \Sigma M_A$$

$$1250(\bar{x}) = \int \underline{\hspace{7cm}}$$

$$\bar{x} = 8 \text{ m} \qquad\qquad \textbf{Ans.}$$

4 - 35. The distributed loadings of soil pressure on the sides and bottom of a retaining wall are shown. Simplify this system to a single resultant force and couple moment acting at A.

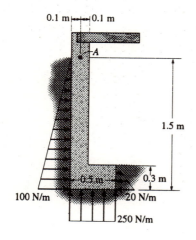

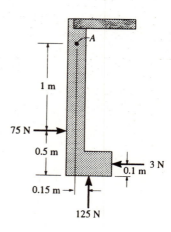

Solution

The resultants of each distributed load are shown on the diagram.

$$\xrightarrow{+} F_{R_x} = \Sigma F_x ; \; F_{R_x} = \underline{\hspace{5cm}} \rightarrow$$

$$+\uparrow F_{R_y} = \Sigma F_y ; \; F_{R_y} = \underline{\hspace{5cm}} \uparrow$$

Thus,

$$F_R = \sqrt{(72)^2 + (125)^2} = 144 \text{ N} \qquad \qquad \textbf{Ans.}$$

$$\theta = \tan^{-1}(\frac{125}{72}) = 60.1° \qquad \qquad \textbf{Ans.}$$

$$\zeta + M_{R_A} = \Sigma M_A ; \; M_{R_A} = \underline{\hspace{5cm}}$$

$$= 89.6 \text{ N} \cdot \text{m} \qquad \qquad \textbf{Ans.}$$

4 - 36. Determine the required intensity w and dimension d of the right support so that the resultant force and couple moment about point A of the system are both zero.

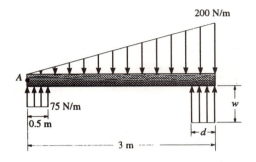

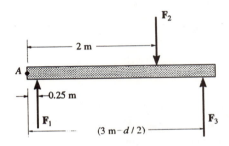

Solution

$$F_1 = \underline{\hspace{6cm}} = 37.5 \text{ N}$$

$$F_2 = \underline{\hspace{6cm}} = 300 \text{ N}$$

$$F_3 = wd$$

$$+\uparrow F_{R_y} = \Sigma F_y ; \quad 0 = \underline{\hspace{7cm}}$$

$$\zeta + M_{R_A} = \Sigma M_A ; \quad 0 = \underline{\hspace{6cm}}$$

Thus,

$$d = 1.5 \text{ m}, \quad w = 175 \text{ N/m} \qquad \textbf{Ans.}$$

5 Equilibrium of a Rigid Body

Equilibrium in Two Dimensions

5 - 1. Compute the horizontal and vertical components of force at pin B. The belt is subjected to a tension of $T = 100$ N and passes over each of the three pulleys.

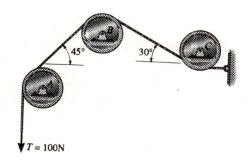

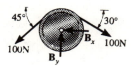

Solution

The free - body diagram of pulley B is drawn first.

$\xrightarrow{+} \Sigma F_x = 0;$ _____

$$B_x = 15.9 \text{ N} \qquad\qquad \textbf{Ans.}$$

$+ \uparrow \Sigma F_y = 0;$ _____

$$B_y = 121 \text{ N} \qquad\qquad \textbf{Ans.}$$

5 - 2. Determine the horizontal and vertical components of reaction at the pin *A* and the reaction at the roller support *B*.

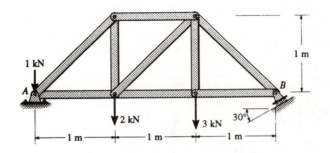

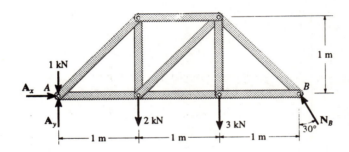

Solution

The free - body diagram of the entire truss is drawn first.

$$\zeta + \Sigma M_A = 0;$$ _____

$$N_B = 3.079 \text{ kN} = 3.08 \text{ kN}$$ **Ans.**

Using this result,

$$\xrightarrow{+} \Sigma F_x = 0;$$ _____

$$A_x = 1.54 \text{ kN}$$ **Ans.**

$$+ \uparrow \Sigma F_y = 0;$$ _____

$$A_y = 3.33 \text{ kN}$$ **Ans.**

5 - 3. The oil rig is supported on the trailer by the pin or axle at A and the frame at B. If the rig has a mass of 50 Mg and a center of gravity at G, determine the force that must be developed along the hydraulic cylinder CD in order to *start* lifting the rig (slowly) off B. Also compute the horizontal and vertical components of reaction at the pin A.

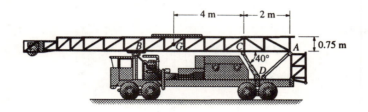

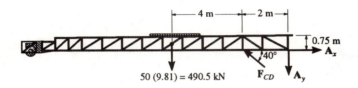

Solution

The free - body diagram of the oil rig is drawn first.

$$(+\ \Sigma M_A = 0;\text{\underline{\hspace{6cm}}}$$

$$F_{CD} = 2289 \text{ kN} = 2.29 \text{ MN} \qquad\qquad \textbf{Ans.}$$

Using this result,

$$\overset{+}{\rightarrow} \Sigma F_x = 0;\text{\underline{\hspace{6cm}}}$$

$$A_x = 1.75 \text{ MN} \qquad\qquad \textbf{Ans.}$$

$$+\uparrow \Sigma F_y = 0;\text{\underline{\hspace{6cm}}}$$

$$A_y = 981 \text{ kN} \qquad\qquad \textbf{Ans.}$$

5 - 4. The boom provides a long - reach capacity by using the telescopic boom segment *DE*. The entire boom is supported by a pin at *A* and by the telescopic hydraulic cylinder *BC*, which can be considered as a two - force member. The rated load capacity of the boom is measured by a maximum force developed in the hydraulic cylinder. If this maximum force is developed when the boom supports a mass *m* = 6 Mg and its length is *l* = 40 m and θ = 60°, determine the greatest mass that can be supported when the boom length is extended to *l* = 50 m and θ = 45°. Neglect the weight of the boom and the size of the pulley at *E*.

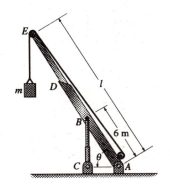
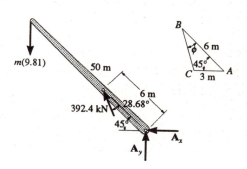

Solution

The free - body diagram of the boom is drawn first.

$$\zeta + \Sigma M_A = 0;\underline{\hspace{10cm}}$$

$$(F_{BC})_{\max} = 392.4 \text{ kN}$$

When θ = 45° the direction of *BC* must be determined from the geometry of the triangle *ABC*.

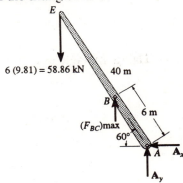

$$BC = \sqrt{(3)^2 + (6)^2 - 2(6)(3)\cos 45°} = 4.421 \text{ m}$$

$$\frac{\sin\phi}{3} = \frac{\sin 45°}{4.421}; \quad \phi = 28.68°$$

The free - body diagram of the boom in this position is shown.

$$\zeta + \Sigma M_A = 0;\underline{\hspace{10cm}}$$

$$m = 3.26 \text{ Mg} \qquad\qquad\qquad \textbf{Ans.}$$

5 - 5. The sports car has a mass of 1.5 Mg and a mass center at G. If the front two springs each have a stiffness of $k_A = 58$ kN/m and the rear two springs each have a stiffness of $k_B = 65$ kN/m, determine their compression when the car is parked on the 30° incline. Also, what frictional force F_B must be applied to each of the rear wheels to hold the car in equilibrium?

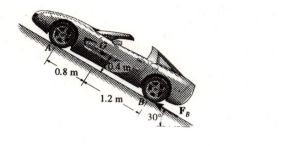
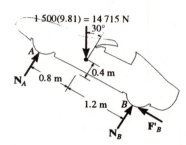

Solution

The compression of the springs can be determined once the normal force at A and B is known. The free - body diagram of the car is drawn first.

$$\xrightarrow{\;+} \Sigma F_x = 0; \underline{\hspace{9cm}}$$

$$F'_B = 7357.5 \text{ N}, \quad F_B = \frac{F'_B}{2} = 3.68 \text{ kN} \qquad\qquad \textbf{Ans.}$$

$$\left(+ \Sigma M_B = 0; \underline{\hspace{9cm}}\right.$$

$$N_A = 6174.6 \text{ N}$$

Using this result,

$$+\nearrow \Sigma F_y = 0; \underline{\hspace{9cm}}$$

$$N_B = 6568.9 \text{ N}$$

Use $F_s = kx_B$ to determine the compression in the springs.

$$x_A = \frac{(\qquad\qquad)}{(\qquad\qquad)} = 53.2 \text{ mm} \qquad\qquad \textbf{Ans.}$$

$$x_B = \frac{(\qquad\qquad)}{(\qquad\qquad)} = 50.5 \text{ mm} \qquad\qquad \textbf{Ans.}$$

5 - 6. Determine the unknown force components acting at the ball - and scoket joint, A, roller B, and cable DC.

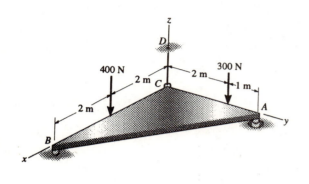
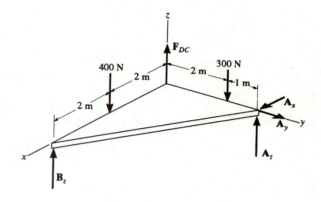

Solution

The free - body diagram of the plate is drawn first.

$$\Sigma F_x = 0; \quad A_x = 0 \qquad \qquad \textbf{Ans.}$$

$$\Sigma F_y = 0; \quad A_y = 0 \qquad \qquad \textbf{Ans.}$$

Use the right - hand rule to write the scalar equations of moment equilibrium.

$$\Sigma M_x = 0; \underline{\hspace{6cm}}$$

$$A_z = 200 \text{ N} \qquad \qquad \textbf{Ans.}$$

$$\Sigma M_y = 0; \underline{\hspace{6cm}}$$

$$B_z = 200 \text{ N} \qquad \qquad \textbf{Ans.}$$

Using these results,

$$\Sigma F_z = 0; \underline{\hspace{6cm}}$$

$$F_{DC} = 300 \text{ N} \qquad \qquad \textbf{Ans.}$$

5 - 7. Determine the x, y, z force components acting at the ball - and - socket joint A and journal bearing B of the shaft.

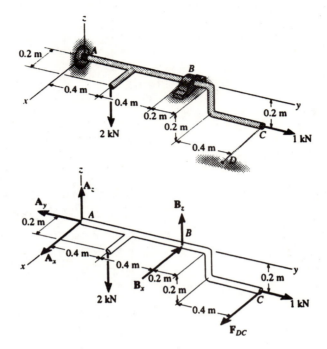

Solution

The free - body diagram of the shaft is drawn first.
Use the right - hand rule to write the scalar equations of moment equilibrium.

$\Sigma M_y = 0;$ _____

$F_{DC} = 2$ kN   **Ans.**

$\Sigma F_y = 0;$ _____

$A_y = 1$ kN   **Ans.**

Use the result for F_{DC}.

$\Sigma M_z = 0;$ _____

$B_x = 3.5$ kN   **Ans.**

Use the results for B_x and F_{DC}.

$\Sigma F_x = 0;$ _____

$A_x = 1.5$ kN   **Ans.**

$\Sigma M_x = 0;$ _____

$B_z = 0.75$ kN   **Ans.**

Use the results for B_z.

$\Sigma F_z = 0;$ _____

$A_z = 1.25$ kN   **Ans.**

5 - 8. The space truss is supported by a ball - and - socket joint at A and short links, two at C and one at D. Determine the x,y,z components of reaction at A and the force in each link.

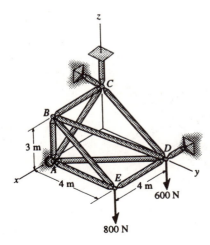

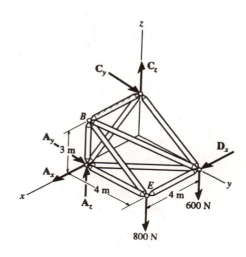

Solution

The free - body diagram of the truss is drawn first.
Use the right - hand rule to write the scalar equations of moment equilibrium.

$\Sigma M_x = 0;$ _____

$\Sigma M_y = 0;$ _____

$\Sigma M_z = 0;$ _____

$\Sigma F_x = 0;$ _____

$\Sigma F_y = 0;$ _____

$\Sigma F_z = 0;$ _____

Thus,

$A_x = -1.87$ kN   **Ans.**

$A_y = 1.87$ kN   **Ans.**

$A_z = 800$ N   **Ans.**

$C_y = -1.87$ kN, $C_z = 600$ N   **Ans**

$D_x = 1.87$ kN   **Ans.**

89

5 - 9. Determine the tension in the supporting cables BC and BD and the components of reaction at the ball - and - socket joint A of the boom. The boom supports a drum having a mass of 200 kg at F. Points C and D lie on the $x - y$ plane.

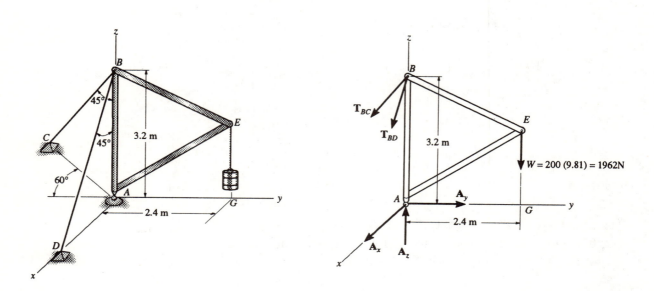

Solution

The free - body diagram of the entire boom is drawn first. Here,

$$\mathbf{T}_{BD} = T_{BD} \sin45°\mathbf{i} - T_{BD} \cos45°\mathbf{k} = 0.7071T_{BD}\mathbf{i} - 0.7071T_{BD}\mathbf{k}$$
$$\mathbf{T}_{BC} = -T_{BC}(\sin45°)(\sin60°)\mathbf{i} - T_{BC}(\sin45°)(\cos60°)\mathbf{j} - T_{BC}(\cos45°)\mathbf{k}$$
$$= -0.6124T_{BC}\mathbf{i} - 0.3536T_{BC}\mathbf{j} - 0.7071T_{BC}\mathbf{k}$$

Apply the equilibrium equations

$\Sigma F_x = 0;$_____

$\Sigma F_y = 0;$_____

$\Sigma F_z = 0;$_____

$\Sigma \mathbf{M}_A = 0; \ \mathbf{r}_{AB} \times (\mathbf{T}_{BC}+\mathbf{T}_{BD}) + \mathbf{r}_{AG} \times \mathbf{W}$

_____ $\times (\mathbf{T}_{BC} + \mathbf{T}_{BD}) +$_____ $\times$_____

Substituting $\mathbf{T}_{BC}$ and $\mathbf{T}_{BD}$, carrying out the cross product, equating the $\mathbf{i}$ and $\mathbf{j}$ components yields

$\Sigma M_x = 0; \ 1.131T_{BC} - 4708.8 = 0$

$\Sigma M_y = 0; \ 2.263T_{BD} - 1.960T_{BC} = 0$

Thus,

$A_x = 0, \ A_y = 1.47$ kN, $A_z = 7.45$ kN   **Ans.**

$T_{BC} = 4.16$ kN, $T_{BD} = 3.60$ kN   **Ans.**

90

6 Structural Analysis

The Method of Joints

6 - 1. Compute the force in each member of the *Warren truss* and indicate whether the members are in tension or compression.

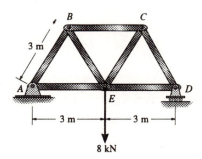

Solution

Due to symmetry, $A_x = 0$, $A_y = D_y = 4$ kN. Start at joint A since there are two unknowns there. First we draw the free - body diagram.

$+\uparrow \Sigma F_y = 0;$ _____

$\xrightarrow{+} \Sigma F_x = 0;$ _____

$F_{AB} = 4.62$ kN (C) $F_{AE} = 2.31$ kN (T)   **Ans.**

Using the result for F_{AB}, the free - body diagram of joint B is drawn.

$+\uparrow \Sigma F_y = 0;$ _____

$\xrightarrow{+} \Sigma F_x = 0;$ _____

$F_{BE} = 4.62$ kN (T) $F_{BC} = 4.62$ kN (C)   **Ans.**

By symmetry :

$F_{DE} = 2.31$ kN (T)   **Ans.**

$F_{DC} = 4.62$ kN (C)   **Ans.**

$F_{EC} = 4.62$ kN (T)   **Ans.**

6 - 2. Determine the force in each member of the truss and indicate whether the members are in tension or compression.

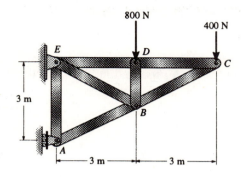

Solution

Here we do not have to determine the support reactions. We begin at joint C
since there is one known force and two unknowns.

$+\uparrow \Sigma F_y = 0;$ _____

$$F_{CB} = 894.4 \text{ N} = 894 \text{ N (C)} \qquad \textbf{Ans.}$$

$\xrightarrow{+} \Sigma F_x = 0;$ _____

$$F_{CD} = 800 \text{ N (T)} \qquad \textbf{Ans.}$$

Joint D :

$\xrightarrow{+} \Sigma F_x = 0; \quad F_{DE} = 800 \text{ N (T)} \qquad \textbf{Ans.}$

$+\uparrow \Sigma F_y = 0; \quad F_{DB} = 800 \text{ N (C)} \qquad \textbf{Ans.}$

Joint B :

$+\nwarrow \Sigma F_y = 0;$ _____

$$F_{BE} = 894.4 \text{ N} = 894 \text{ N (T)} \qquad \textbf{Ans.}$$

$\nearrow^{+} \Sigma F_x = 0;$ _____

$$F_{BA} = 1788.9 \text{ N} = 1.79 \text{ kN (C)} \qquad \textbf{Ans.}$$

Joint A :

$+\uparrow \Sigma F_y = 0;$ _____

$$F_{AE} = 800 \text{ N (T)} \qquad \textbf{Ans.}$$

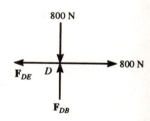

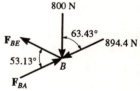

6 - 3. A sign is subjected to a wind loading that exerts horizontal forces of 3 kN on joints *B* and *C* of one of the side supporting trusses. Determine the force in memders *CB*, *CD*, *DB*, and *DE* of the truss and state whether the members are in tension or compression. *Note* : Analyze joint *C*, then joint *D*.

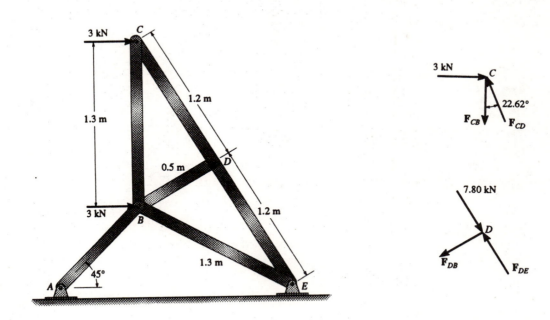

Solution

Joint *C*

$$\xrightarrow{+}\ \Sigma F_x = 0;$$ ——————————————————

$$F_{CD} = 7.80 \text{ kN (C)}$$ **Ans.**

$$+\uparrow \Sigma F_y = 0;$$ ——————————————————

$$F_{CB} = 7.20 \text{ kN (T)}$$ **Ans.**

Joint *D*

$$\nearrow^+\ \Sigma F_x = 0; \quad F_{DB} = 0$$ **Ans.**

$$+\nwarrow \Sigma F_y = 0;$$ ——————————————————

$$F_{DE} = 7.80 \text{ N (C)}$$ **Ans.**

The Method of Sections

6 - 4. The *Pratt bridge truss* is subjected to the loading shown. Determine the force in members *CL*, *ML*, and *CD*, and indicate whether these members are in tension or compression.

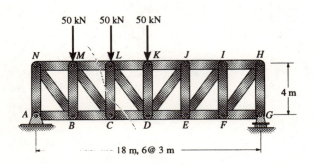

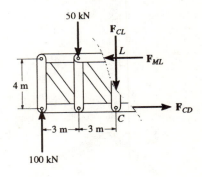

Solution

Before the truss is sectioned the reaction at A is determined to be $A_x = 0$, $A_y = 100$ kN. The free-body diagram of the left portion of the sectioned truss is then drawn.

Each force can be determined directly from a single equation of equilibrium.

$$+ \uparrow \Sigma F_y = 0; \underline{\hspace{8cm}}$$

$$F_{CL} = 50 \text{ kN (C)} \qquad\qquad \textbf{Ans.}$$

$$\curvearrowleft + \Sigma M_C = 0; \underline{\hspace{7cm}}$$

$$F_{ML} = 112.5 \text{ kN (C)} \qquad\qquad \textbf{Ans.}$$

$$\curvearrowleft + \Sigma M_L = 0; \underline{\hspace{7cm}}$$

$$F_{CD} = 112.5 \text{ kN (T)} \qquad\qquad \textbf{Ans.}$$

6 - 5. Determine the force in members *CM, CB,* and *LM* of *the* truss and indicate whether the members are in tension or compression.

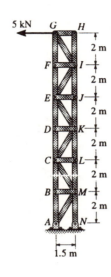

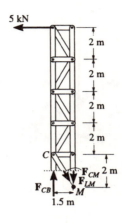

Solution

The free - body diagram of the top portion of the sectioned truss can be used to determine the required member forces.

Each of these forces can be determined directly from a single equation of equilibrium.

$$\xrightarrow{+} \Sigma F_x = 0; \underline{\hspace{10cm}}$$

$$F_{CM} = 833 \text{ kN (T)} \qquad\qquad \textbf{Ans.}$$

$$\left(+ \Sigma M_M = 0; \underline{\hspace{8cm}} \right.$$

$$F_{CB} = 33.3 \text{ kN (C)} \qquad\qquad \textbf{Ans.}$$

$$\left(+ \Sigma M_C = 0; \underline{\hspace{8cm}} \right.$$

$$F_{LM} = 26.7 \text{ kN (T)} \qquad\qquad \textbf{Ans.}$$

6 - 6. Determine the force in members *BF*, *BC*, and *GF* of the *Fink truss* and indicate whether the members are in tension or compression.

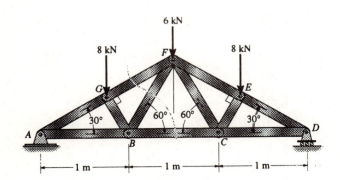
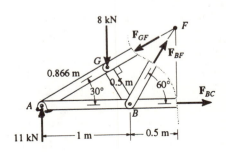

Solution

The reaction at *A* is $A_x = 0$, $A_y = 11$ kN. The free - body diagram of the left portion of the sectioned truss will be considered.

Each of these forces can be determined directly from a single equation of equilibrium.

$\zeta + \Sigma M_A = 0;$ _____

$$F_{BF} = 6.93 \text{ kN (T)} \qquad\qquad \textbf{Ans.}$$

$\zeta + \Sigma M_F = 0;$ _____

$$F_{BC} = 12.1 \text{ kN (T)} \qquad\qquad \textbf{Ans.}$$

$\zeta + \Sigma M_B = 0;$ _____

$$F_{GF} = 18.0 \text{ kN (C)} \qquad\qquad \textbf{Ans.}$$

6 - 7. Determine the force in members *CF*, *CD*, and *GF* of the roof truss and indicate whether the members are in tension or compression.

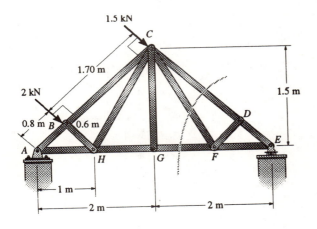

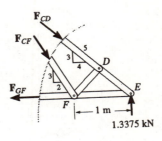

Solution

Show that the reaction at *E* is E_y = 1.3375 kN by summing moments about point *A* on the entire truss. The free - body diagram of the right portion of the sectioned truss will be considered.

$$\curvearrowleft + \Sigma M_E = 0; \quad F_{CF} = \underline{\hspace{3cm}} \qquad \text{Ans.}$$

$$+ \uparrow \Sigma F_y = 0; \underline{\hspace{8cm}}$$

$$F_{CD} = 2.23 \text{ kN (C)} \qquad \text{Ans}$$

Using this result,

$$\xrightarrow{+} \Sigma F_x = 0; \underline{\hspace{8cm}}$$

$$F_{GF} = 1.78 \text{ kN (T)} \qquad \text{Ans.}$$

6 - 8. The *Warren truss* is used to support a staircase. Dtermine the force in members *DF*, *CE*, and *ED*, and state whether the members are in tension or compression. Assume all joints are pinned.

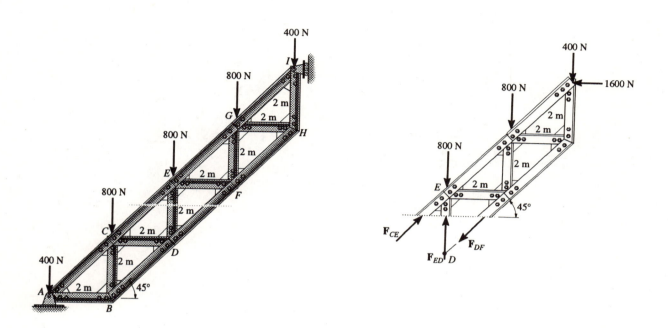

Solution

Before the truss is sectioned the horizontal reaction of *I* is determined to be I_x = 1600 N. The free - body diagram of the top portion of the section will be considered.

$$\left(+ \Sigma M_E = 0; \underline{\hspace{10cm}}\right.$$

$$F_{DF} = 2262.7 \text{ N} = 2.26 \text{ kN (T)} \qquad\qquad \textbf{Ans.}$$

$$\left(+ \Sigma M_D = 0; \underline{\hspace{10cm}}\right.$$

$$F_{CE} = 4525.5 \text{ N} = 4.53 \text{ kN(C)} \qquad\qquad \textbf{Ans.}$$

$$+ \uparrow \Sigma F_y = 0; \underline{\hspace{10cm}}$$

$$F_{ED} = 400 \text{ N (C)} \qquad\qquad \textbf{Ans.}$$

Frames and Machines

6 - 9. The principles of a *differential chain block* are indicated schematically in the figure. Determine the magnitude of the force **P** needed to support the 800 - N force. Also, compute the distance x where the cable must be attached to bar AB so the bar remains horizontal. All pulleys have a radius of 60 mm.

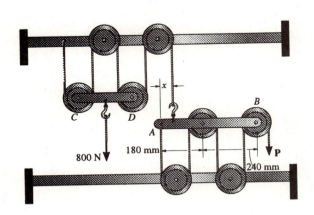
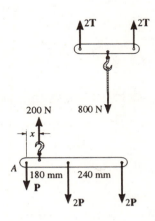

Solution

From the free - body diagram of the member CD we can obtain the force in the top cable.

$$+\uparrow \Sigma F_y = 0; \underline{\hspace{10cm}}$$

$$T = 200 \text{ N}$$

Now consider the free - body diagram of the member AB.

$$+\uparrow \Sigma F_y = 0; \underline{\hspace{10cm}}$$

$$P = 40 \text{ N} \hspace{6cm} \textbf{Ans.}$$

Using this result,

$$\left(+ \Sigma M_A = 0; \underline{\hspace{9cm}} \right.$$

$$x = 240 \text{ mm} \hspace{6cm} \textbf{Ans.}$$

6 - 10. Determine the horizontal and vertical components of force at pins A and C of the two - member frame.

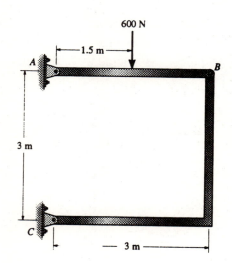

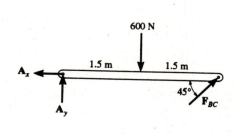

Solution

Note that CB is a two - force member. The solution can be determined from the free - body diagram of member AB.

$\zeta + \Sigma M_A = 0;$ _____

$$F_{BC} = 424.26 \text{ N}$$

Using this result,

$\xrightarrow{+} \Sigma F_x = 0;$ _____

$$A_x = 300 \text{ N} \hspace{3cm} \textbf{Ans.}$$

$+ \uparrow \Sigma F_y = 0;$ _____

$$A_y = 300 \text{ N} \hspace{3cm} \textbf{Ans.}$$

$$C_x = C_y = 424.26(0.707) = 300 \text{ N} \hspace{2cm} \textbf{Ans.}$$

6 - 11. Determine the horizontal and vertical componenets of force at pin A of the two - member frame.

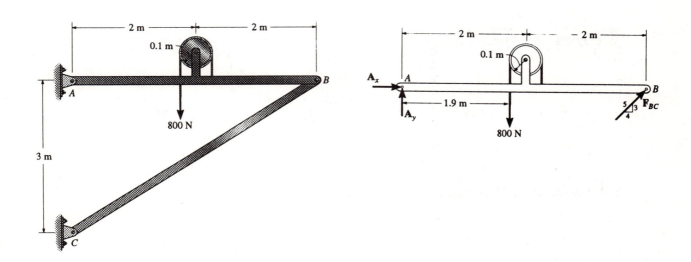

Solution

Here BC is a two - force member. Consider the free - body diagram of member AB and attached pulley.

$$\left(+\; \Sigma M_A = 0;\underline{\hspace{6cm}}\right.$$

$$F_{BC} = 633.3 \text{ N}$$

Using this result,

$$\xrightarrow{+}\; \Sigma F_x = 0;\underline{\hspace{6cm}}$$

$$A_x = 507 \text{ N} \hspace{4cm} \textbf{Ans.}$$

$$+\uparrow \Sigma F_y = 0;\underline{\hspace{6cm}}$$

$$A_y = 420 \text{ N} \hspace{4cm} \textbf{Ans.}$$

6 - 12. Determine the horizontal and vertical components of force at pins D and E of the four - member frame.

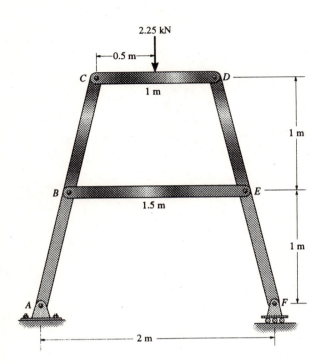

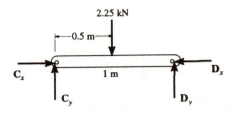

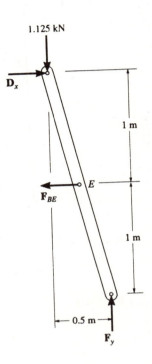

Solution

Start with the free - body diagram of member CD to obtain D_y.

$$\zeta + \Sigma M_C = 0; \underline{\hspace{10cm}}$$
$$D_y = 1.125 \text{ kN} = 1.12 \text{ kN} \qquad\qquad\qquad \textbf{Ans.}$$

By inspection note that BE is a two - force member. Consider the free - body diagram of member DEF.

$$+\uparrow \Sigma F_y = 0; \underline{\hspace{7cm}}$$
$$F_y = 1.125 \text{ kN}$$

Using this result,

$$\zeta + \Sigma M_D = 0; \underline{\hspace{9cm}}$$

$$\overset{+}{\rightarrow} \Sigma F_x = 0; \underline{\hspace{8cm}}$$

Thus,

$$F_{BE} = 562 \text{ N}, \quad D_x = 562 \text{ N} \qquad\qquad \textbf{Ans.}$$

Because BE is a two - force member,

$$E_x = 562 \text{ N}, \quad E_y = 0 \qquad\qquad \textbf{Ans.}$$

6 - 13. The floor beams AB and BC are stiffened using the two tie rods CD and AD. Determine the force along each rod. Assume the three contacting members at B are smooth, the thickness of the members can be neglected, and the joints at A, C, and D are pins.

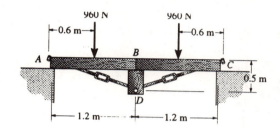

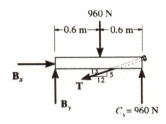

Solution

The support reaction C_y can be determined by considering the entire framework, $C_y = 960$ N.

By inspection AD, CD, and BD are two - force members. Consider the free - body diagram of member BC.

$$\zeta + \Sigma M_B = 0;\underline{\hspace{8cm}}$$

$$T = 1.25 \text{ kN} \qquad\qquad \textbf{Ans.}$$

6 - 14. The hoist supports the 125 - kg engine. Detrmine the force the load creates in member *FB*, which contains the hydraulic cylinder *H*, and in member *FB*.

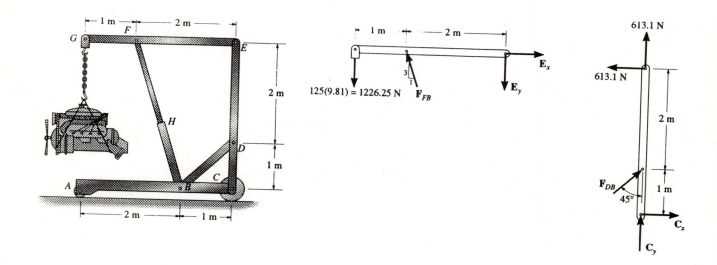

Solution

By investigation *FB* and *BD* are two - force members. Using the free - body diagram of *GFE*, we have

$$\zeta + \Sigma M_E = 0; \underline{\hspace{5cm}}$$

$$F_{FB} = 1938.9 \text{ N} = 1.94 \text{ kN} \qquad\qquad \textbf{Ans.}$$

Using this result,

$$\xrightarrow{+} \Sigma F_x = 0; \underline{\hspace{5cm}}$$

$$E_x = 613.1 \text{ N}$$

$$+ \uparrow \Sigma F_y = 0; \underline{\hspace{5cm}}$$

$$E_y = 613.1 \text{ N}$$

The free - body diagram of member *EDC* is now considered to find the force in *DB*.

$$\zeta + \Sigma M_C = 0; \underline{\hspace{5cm}}$$

$$F_{DB} = 2.60 \text{ kN} \qquad\qquad \textbf{Ans.}$$

6-15. The scissors lift consists of *two* hydraulic cylinders, *DE*, symmetrically located on *each side* of the platform. The platform has a uniform mass of 60 kg, with a center of gravity at G_1. The load of 85 kg, with center of gravity at G_2, is centrally located on each side of the platform. Determine the force in each of the hydraulic cylinders for equilibrium. Rollers are located at *B* and *D*.

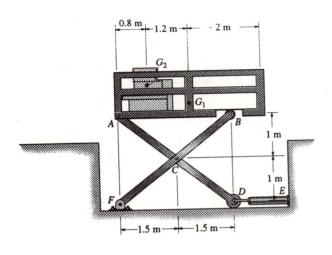

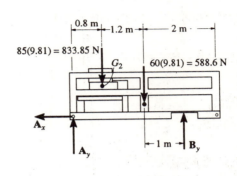

Solution

Platform

$\zeta + \Sigma M_A = 0;$ _____

$$B_y = 614.8 \text{ N}$$

$\xrightarrow{+} \Sigma F_x = 0; \ A_x = 0$

$+\uparrow \Sigma F_y = 0;$ _____

$$A_y = 807.7 \text{ N}$$

Members *BCF* and *ACD*

$\zeta + \Sigma M_D = 0;$ _____

$\zeta + \Sigma M_F = 0;$ _____

$$C_x = 2133.8 \text{ N}; \ C_y = 192.9 \text{ N}$$

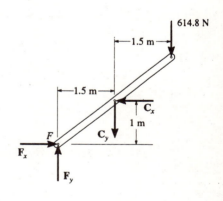

Member *ACD*

$\xrightarrow{+} \Sigma F_x = 0;$ _____

$$F_{DE} = 2133.7 \text{ N}$$

For one cylinder,

$$F'_{DE} = \frac{2133.7}{2} = 1.07 \text{ kN} \qquad\qquad \textbf{Ans.}$$

7 Internal Forces

Internal Forces Developed in Structural Members

7 - 1. The axial forces act on the shaft as shown. Determine the internal normal force at points A and B.

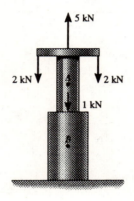
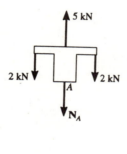
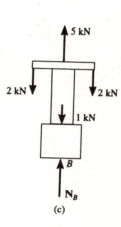

(c)

Solution

Due to the external axial loading, no shear or moment will act on the section through A as shown on the free - body diagram of the top segment.

$$+\uparrow \Sigma F_z = 0; \underline{\hspace{10cm}}$$

$$N_A = 1 \text{ kN} \qquad\qquad \textbf{Ans.}$$

Using the free - body diagram of the top segment for section B,

$$+\uparrow \Sigma F_z = 0; \underline{\hspace{10cm}}$$

$$N_B = 0 \qquad\qquad \textbf{Ans.}$$

7-2. If the members are in smooth contact with one another at *A*, *B*, and *C* with no fasteners, determine the shear force developed at a horizontal section through point *D* of the support.

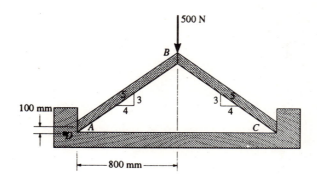

Solution

By inspection, *AB* and *BC* are two-force members. The force in these members can be determined from the free-body diagram of joint *B* .

$$+\uparrow \Sigma F_y = 0;\ \underline{\hspace{9cm}}$$

$$F = 416.7\ \text{N}$$

Use the portion of the section about point *D*.

$$\stackrel{+}{\rightarrow} \Sigma F_x = 0;\ \underline{\hspace{8cm}}$$

$$V_D = 333\ \text{N} \qquad\qquad\qquad \textbf{Ans.}$$

7 - 3. Determine the internal normal force, shear force, and moment at point E of the oleo strut AB of the aircraft landing gear.

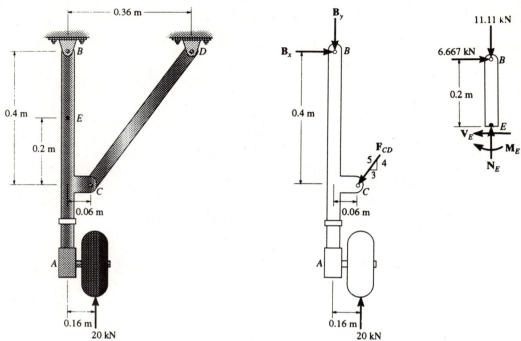

Solution

Here we will consider segment BE, but first we must determine the reactions at B. Member DC is a two - force member. The free - body diagram of the oleo strut must be used.

$$\zeta + \Sigma M_B = 0; \underline{\hspace{5cm}}$$

$$F_{CD} = 11.11 \text{ kN}$$

$$\xrightarrow{+} \Sigma F_x = 0; \underline{\hspace{5cm}} \qquad B_x = 6.667 \text{ kN}$$

$$+\uparrow \Sigma F_y = 0; \underline{\hspace{5cm}} \qquad B_y = 11.11 \text{ kN}$$

Using the free - body diagram of segment BE, we have

$$\xrightarrow{+} \Sigma F_x = 0; \underline{\hspace{5cm}} \qquad V_E = 6.67 \text{ kN} \qquad \textbf{Ans.}$$

$$+\uparrow \Sigma F_y = 0; \underline{\hspace{5cm}} \qquad N_E = 11.1 \text{ kN} \qquad \textbf{Ans.}$$

$$\zeta + \Sigma M_E = 0; \underline{\hspace{5cm}} \qquad M_E = 1.33 \text{ kN} \cdot \text{m} \qquad \textbf{Ans.}$$

7 - 4. Determine the internal normal force, shear force, and moment at point F of the frame.

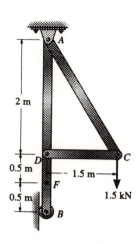

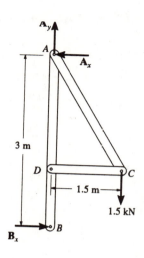

Solution

Segment FB will be used and so we must first determine the reaction at B from the free - body diagram of the entire frame.

$$\zeta + \Sigma M_A = 0; \quad \underline{\hspace{10cm}}$$

$$B_x = 0.750 \text{ kN}$$

Using the free - body diagram of segment BF :

$$\overset{+}{\rightarrow} \Sigma F_x = 0; \quad \underline{\hspace{6cm}}$$

$$V_F = 0.750 \text{ kN} \qquad \qquad \textbf{Ans.}$$

$$+ \uparrow \Sigma F_y = 0; \quad N_F = \underline{\hspace{5cm}} \qquad \textbf{Ans.}$$

$$\zeta + \Sigma M_F = 0; \quad \underline{\hspace{6cm}}$$

$$M_F = 0.750 \text{ kN} \qquad \qquad \textbf{Ans.}$$

8 Friction

Problems Involving Dry Friction

8-1. A 17-kg ladder has a center of mass at G. If the coefficients of static friction at A and B are $\mu_A = 0.3$ and $\mu_B = 0.2$. respectively, determine the smallest horizontal force that must be applied at point C in order to push the ladder forward.

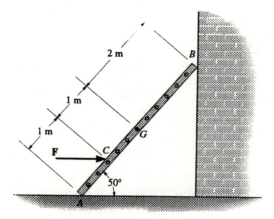 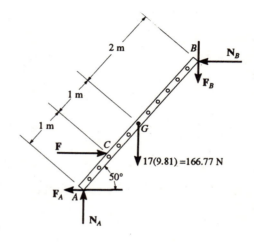

Solution

The free-body diagram of the ladder is drawn first. There are 5 unknowns, for which 3 equations of equilibrium and 2 frictional equations are needed for the solution. This impending motion occurs at at A and B.
Thus,

$$F_A = 0.3N_A, \quad F_B = 0.2N_B$$

$\xrightarrow{+} \Sigma F_x = 0;$ _____

$+\uparrow \Sigma F_y = 0;$ _____

$\big(+ \Sigma M_A = 0;$ _____

Thus,

$$N_A = 195.9 \text{ N}, \ N_B = 145.4 \text{ N}$$

$$F = 204 \text{ N}$$

Ans.

8 - 2. A uniform beam has a mass of 18 kg and rests on two surfaces at A and B for which the coefficient of static friction is $\mu_s = 0.2$. Determine the maximum distance x to which the vertical force of 500 N can be placed on the beam without causing the beam to slip. Neglect the thickness of the beam.

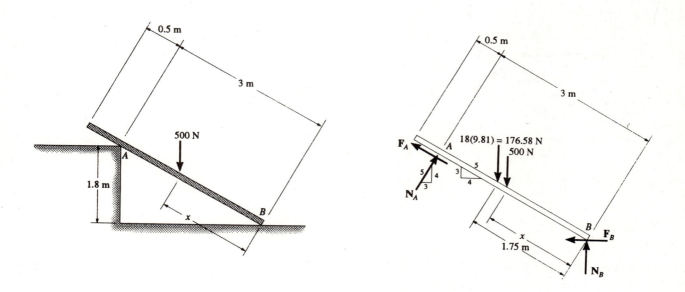

Solution

The free - body diagram has 5 unknowns for which there are 3 equations of equilibrium and 2 equations of friction available for the solution. Since impending motion must occur at both A and B,

$$F_A = 0.2N_A$$

$$F_B = 0.2N_B$$

$\xrightarrow{+} \Sigma F_x = 0;$ _____

$+ \uparrow \Sigma F_y = 0;$ _____

$\zeta + \Sigma M_B = 0;$ _____

Thus,

$$N_A = 216.8 \text{ N}, N_B = 477.1 \text{ N},$$

$$x = 1.01 \text{ m} \qquad \qquad \textbf{Ans.}$$

111

8 - 3. The refrigerator has a mass of 200 kg and a center of gravity at G. Determine the force **P** required to move it. Take $\mu_s = 0.4$.

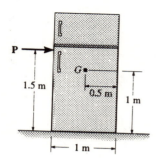

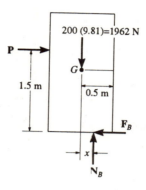

Solution

There are 4 unknowns on the free - body diagram and 3 available equilibrium equations. It is possible for the refrigerator to slip or tip. If we assume tipping then $x = 0.5$ m.

$$\zeta + \Sigma M_B = 0;\ \underline{\hspace{5cm}}$$

$$P = 654 \text{ N} \qquad\qquad \textbf{Ans.}$$

$$+\uparrow \Sigma F_y = 0;\ \underline{\hspace{5cm}}$$

$$N_B = 1962 \text{ N}$$

$$\xrightarrow{+}\ \Sigma F_x = 0;\quad F_B = \underline{\hspace{5cm}}$$

Check to be sure the refrigerator does not slip.

$$(F_B)_{\max} = \underline{\hspace{3cm}} = 785 \text{ N} > 654 \text{ N}\quad \text{OK}$$

8 - 4. The block at D has a mass of 50 kg and rests on a plank at the position shown. The plank is pin - supported at A and rests on a post at B. Neglecting the weight of the plank and post, determine the magnitude of force **P** at E that must be applied in order to pull out the post. The coefficeients of static friction are $\mu_B = 0.3$ and $\mu_C = 0.8$.

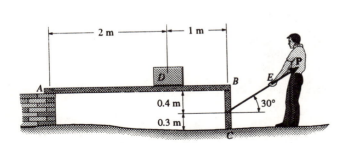

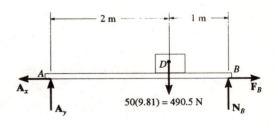

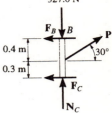

Solution

The free - body diagrams of the plank and the post are first constructed.
We can determine the normal force at B by considering the free - body diagram of the plank.

$$\zeta + \Sigma M_A = 0; \underline{\hspace{6cm}}$$
$$N_B = 327.0 \text{ N}$$

Since $\mathbf{A}_x$ and $\mathbf{A}_y$ are of no interest for the problem solution, consider the free - body diagram of the post which has 4 unknowns.

$$+ \uparrow F_y = 0; \underline{\hspace{8cm}}$$

$$\overset{+}{\rightarrow} \Sigma F_x = 0; \underline{\hspace{9cm}}$$

$$\zeta + \Sigma M_C = 0; \underline{\hspace{8cm}}$$

We need one more equation. Slipping can occur either at B or C. Assume slipping at B.

$$F_B = 0.3(327.0) = 98.1 \text{ N}$$

Soliving :
$$P = 264 \text{ N}$$
$$F_C = 130.8 \text{ N} \qquad\qquad \textbf{Ans.}$$
$$N_C = 195 \text{ N}$$

Check to be sure slipping does not occur at C.

$$(F_C)_{max} = \underline{\hspace{4cm}} = 156 \text{ N} > 130.8 \text{ N} \quad \text{OK}$$

8 - 5. Determine the minimum force **P** needed to push the two 75 - kg cylinders up the incline. The force acts parallel to the plane and the coefficient of static friction at the contracting surfaces are $\mu_A = 0.3$, $\mu_B = 0.25$, and $\mu_C = 0.4$. Each cylinder has a radius of 150 mm.

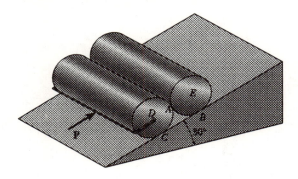

Solution
The free - body diagram of each cylinder indicates there are 7 unknowns.
Cylinder D :

$$\xrightarrow{+}\Sigma F_x = 0; \underline{\hspace{6cm}} \quad (1)$$

$$\nwarrow^+ \Sigma F_y = 0; \underline{\hspace{6cm}} \quad (2)$$

$$\zeta + \Sigma M_D = 0; \underline{\hspace{6cm}} \quad (3)$$

Cylinder E :

$$\xrightarrow{+}\Sigma F_x = 0; \underline{\hspace{6cm}} \quad (4)$$

$$\nwarrow^+ \Sigma F_y = 0; \underline{\hspace{6cm}} \quad (5)$$

$$\zeta + \Sigma M_E = 0; \underline{\hspace{6cm}} \quad (6)$$

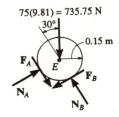

We need one more equation. Slipping can occur at A, B, or C.
Assume the cylinders slip at A, then

$$F_A = 0.3 N_A \quad (7)$$

Solving Eqs. (1) - (7),

$$N_A = 525.5 \text{ N}; \quad F_A = F_B = 157.7 \text{ N}; \quad N_B = 794.8 \text{ N}$$
$$F_C = 157.7 \text{ N}; \quad N_C = 479.5 \text{ N}; \quad P = 1.05 \text{ kN}$$

Now we must check to be sure slipping does not occur at C or B.

$$(F_B)_{max} = \underline{\hspace{4cm}} = 198.7 \text{ N} > 157.7 \text{ N} \quad \text{OK}$$

$$(F_C)_{max} = \underline{\hspace{4cm}} = 191.8 \text{ N} > 157.7 \text{ N} \quad \text{OK}$$

Both cylinders roll, with slipping at A initially assumed. Thus,

$$P = 1.05 \text{ kN} \qquad \qquad \textbf{Ans.}$$

9 Center of Gravity and Centroid

Center of Gravity and Centroid by Integration

9 - 1. Locate the centroid of the parabolic area.

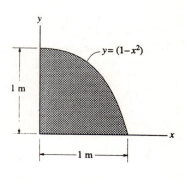

Solution

Using the element of thickness dx as shown express each of the following in terms of x.

$dA = $ _____

$\tilde{x} = $ _____

$\tilde{y} = $ _____

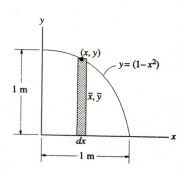

Using the element of thickness dy as shown, express each of the following in terms of y.

$dA = $ _____

$\tilde{x} = $ _____

$\tilde{y} = $ _____

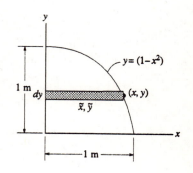

Either set of data can be substituted into the following equations to obtain the solution.

$$\bar{x} = \frac{\int \tilde{x} dA}{\int dA} = \frac{3}{8}\,\text{m} \qquad \textbf{Ans.}$$

$$\bar{y} = \frac{\int \tilde{y} dA}{\int dA} = \frac{2}{5}\,\text{m} \qquad \textbf{Ans.}$$

115

9 - 2. Locate the centroid of the exparabolic segment of area.

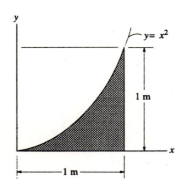 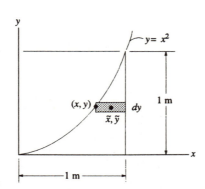 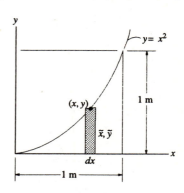

Solution

Using the element of thickness dx as shown express each of the following in terms of x.

$dA = $ _____

$\tilde{x} = $ _____

$\tilde{y} = $ _____

Using the element of thickness dy as shown, express each of the following in terms of y.

$dA = $ _____

$\tilde{x} = $ _____

$\tilde{y} = $ _____

Either set of data can be substituted into the following equations to obtain the solution.

$$\bar{x} = \frac{\int \tilde{x}\,dA}{\int dA} = \frac{3}{4}m \qquad \textbf{Ans.}$$

$$\bar{y} = \frac{\int \tilde{y}\,dA}{\int dA} = \frac{3}{10}m \qquad \textbf{Ans.}$$

116

9 - 3. Locate the centroid of the shaded area.

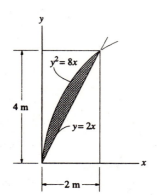 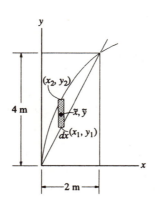 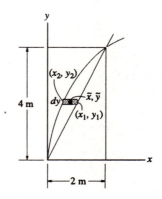

Solution

Using the element of thickness dx as shown express each of the following in terms of x.

$$dA = \text{_____}$$

$$\tilde{x} = \text{_____}$$

$$\tilde{y} = \text{_____}$$

Either set of data can be substituted into the following equations to obtain the solution.

$$\bar{x} = \frac{\int \tilde{x}dA}{\int dA} = 0.8 \text{ m} \qquad \textbf{Ans.}$$

$$\bar{y} = \frac{\int \tilde{y}dA}{\int dA} = 2 \text{ m} \qquad \textbf{Ans.}$$

9 - 4. Locate the center of gravity of the volume generated by revolving the shaded area about the z axis. The material is homogeneous.

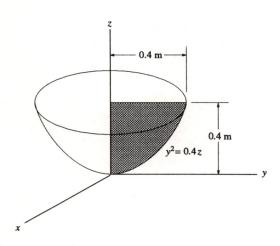 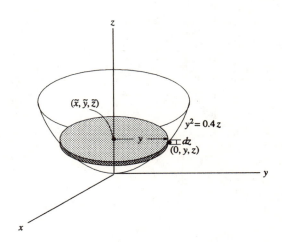

Solution

$$\bar{x} = 0 \qquad \bar{y} = 0 \qquad \text{(By symmetry)} \qquad \textbf{Ans.}$$

Using the disk element of thickness dz shown, express each of the following in terms of z.

$$dV = \underline{\hspace{5cm}}$$

$$\tilde{z} = \underline{\hspace{5cm}}$$

$$\bar{z} = \frac{\int_0^4 z(\pi)(0.4z)\,dz}{\int_0^4 \pi(0.4z)\,dz}$$

$$\bar{z} = 0.267 \text{ m} \qquad \textbf{Ans.}$$

9 - 5. Locate the center of gravity of the homogeneous "bell - shaped" volume formed by revolving the shaded area about the y axis.

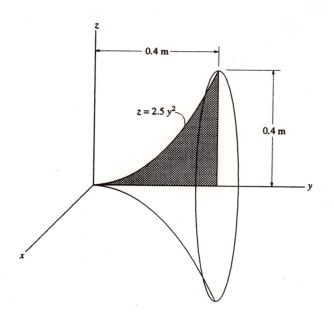
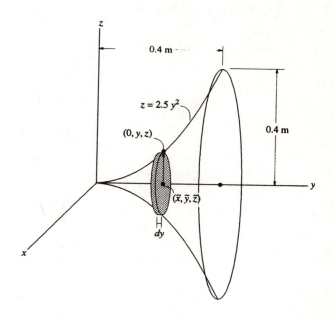

Solution

$$\bar{x} = 0 \qquad \bar{z} = 0 \qquad \text{(By symmetry)} \qquad \text{Ans.}$$

Using the disk element of thickness dx shown, express each of the following in terms of z.

$dV =$ _____

$\tilde{y} =$ _____

$$\bar{y} = \frac{6.25\pi \int_0^4 y^5 dy}{6.25\pi \int_0^4 y^4 dy}$$

$$\bar{y} = 0.333 \text{ m} \qquad\qquad \text{Ans.}$$

9 - 6. The truss is made from three members, each having a mass 6 kg/m. Locate the position $(\bar{x}, \bar{y})$ of the center of mass.

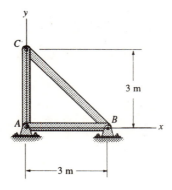

Solution

Since each member has the same mass per length, the centroid will coincide with the center of mass. Locate the centroid of each member relative to the x, y axes and apply the following equations.

$$\bar{x} = \frac{\Sigma \tilde{x} L}{\Sigma L} = \underline{\hspace{8cm}}$$

$$\bar{x} = 1.06 \text{ m} \qquad \qquad \textbf{Ans.}$$

$$\bar{y} = \frac{\Sigma \tilde{y} L}{\Sigma L} = \underline{\hspace{8cm}}$$

$$\bar{y} = 1.06 \text{ m} \qquad \qquad \textbf{Ans.}$$

9 - 7. Determine the distance $\bar{y}$ measured from the x axis to the centroid axis $\bar{x}$ of the beam's cross - sectional area.

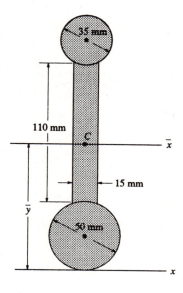

Solution

The cross section is divided into the three composite parts shown. Determine the centroid $\tilde{y}$ to each part from the x axis and apply the equation.

$$\bar{y} = \frac{\Sigma \tilde{y} A}{\Sigma A} = \underline{\hspace{7cm}}$$

$$\bar{y} = 85.9 \text{ mm} \qquad\qquad \textbf{Ans.}$$

9 - 8. Determine the distance $\bar{y}$ measured from the x axis to the centroidal axis of the beam's cross - sectional area.

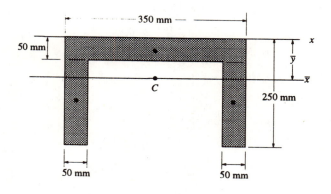

Solution

The shape is divided into the three composite parts shown. Determine the centroid $\tilde{y}$ of each part measured from the x axis and apply the following equation.

$$\bar{y} = \frac{\Sigma \tilde{y} A}{\Sigma A} = \underline{\hspace{8cm}}$$

$$\bar{y} = 91.7 \text{ mm} \qquad\qquad\qquad \textbf{Ans.}$$

Theorems of Pappus and Guldinus

9 - 9. Determine the volume and surface area of the circular segment. Exclude the area of the cross - sections at the ends.

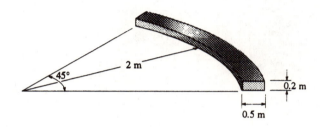

Solution

$$V = \theta\, \bar{x}\, A = \underline{\hspace{5cm}}$$

$$V = 0.177 \text{ m}^2 \qquad\qquad\qquad \textbf{Ans.}$$

$$A = \Sigma\theta\, \bar{x}\, L = \underline{\hspace{5cm}}$$

$$A = 2.47 \text{ m}^2 \qquad\qquad\qquad \textbf{Ans.}$$

9-10. Determine the surface area (excluding the base) and volume of the tank.

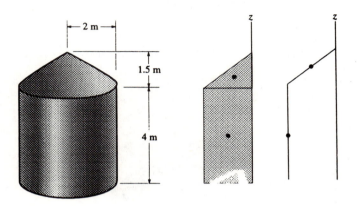

Solution

The surface area is formed by revolving the two line segments having centroids indicated by the dots.

$$A = \Sigma \theta \bar{x} L = \underline{\hspace{5cm}}$$

$$A = 66.0 \text{ m}^2 \qquad \text{Ans.}$$

The volume is formed by revolving the two area segments having centroids indicated by the dots.

$$V = \Sigma \theta \bar{x} A = \underline{\hspace{5cm}}$$

$$V = 56.5 \text{ m}^3 \qquad \text{Ans.}$$

10 Moments of Inertia of an Area

Moment of Inertia for an Area by Integration

10 - 1. The irregular area has a moment of inertia about the A - A axis of $35(10^6)$ mm^4. If the total area is $12.0(10^3)$ mm^2, determine the moment of inertia of the area about the B - B axis. The D - D axis passes through the centroid C of the area.

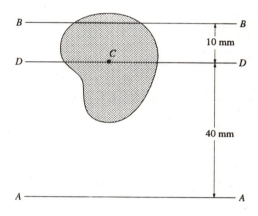

Solution

The parallel - axis theorem must be applied twice to obtain I_{BB} since it depends upon knowing the centroidal moment of inertia.

$$I_{AA} = I_{DD} + A(40)^2$$

_____

$$I_{DD} = 15.8(10^6) \text{ mm}^4$$

$I_{BB} = $ _____

$$I_{BB} = 17.0(10^6) \text{ mm}^4 \qquad \textbf{Ans.}$$

10 - 2. Determine the moment of inertia of the parabolic area about the x axis.

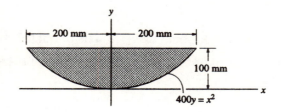

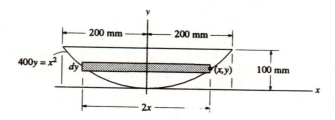

Solution

Use the differential element of thickness dy,

$$I_x = \int y^2 dA = \int_0^{100}$$ _____

$$I_x = 114(10^{-6}) \text{ m}^4 \qquad\qquad\qquad \textbf{Ans.}$$

10 - 3. Determine the radius of gyration k_y of the parabolic area.

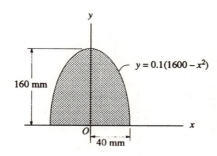

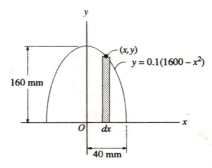

Solution

To find k_y we must first determine I_y and A.

$$A = \int_A dA = \int_{-40}^{40} \underline{\hspace{6cm}}$$

$$A = 8533.3 \text{ mm}^2$$

$$I_y = \int_A x^2 \, dA = \int_{-40}^{40} \underline{\hspace{6cm}}$$

$$I_y = 2730.67(10^3) \text{ mm}^4$$

$$k_y = \sqrt{\frac{I_y}{A}} = \sqrt{\frac{2730.67(10^3)}{8533.3}} = 17.9 \text{ mm} \qquad \textbf{Ans.}$$

10 - 4. Determine the moment of inertia of the area about the x' axis. Then, using the parallel - axis theorem, compute the moment of inertia about the $\bar{x}$ axis that passes through the centroid C of the area. $\bar{y} = 120$mm.

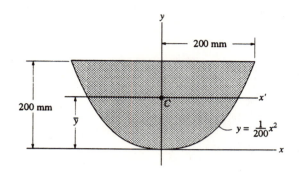

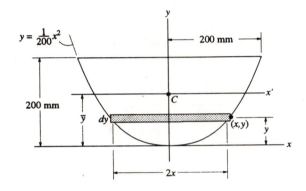

Solution

$$I_x = \int y^2 dA = \int_0^{200} \underline{\hspace{4cm}}$$

$$I_x = 914(10^6) \text{ mm}^4 \qquad\qquad \textbf{Ans.}$$

$$A = \int_0^{200} \underline{\hspace{5cm}}$$

$$A = 53.33(10^3) \text{ mm}^2$$

$$I_x = \bar{I}_{x'} + Ad^2$$

$$\underline{\hspace{7cm}}$$

$$\bar{I}_{x'} = 146(10^6) \text{ mm}^4 \qquad\qquad \textbf{Ans.}$$

Moment of Inertia for a Composite Area

10 - 5. Determine the location $\bar{y}$ of the centroid C of the beam's cross-sectional area. Then compute the moment of inertia of the area about the x' axis.

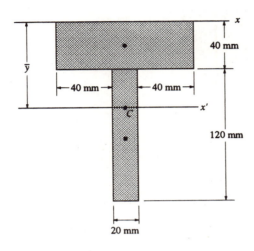

Solution

The shape is divided into two composite parts. Determine the centroid $\tilde{y}$ to each part measured from the x' axis and apply the equation.

$$\bar{y} = \frac{\Sigma \tilde{y} A}{\Sigma A} = \underline{\hspace{8cm}}$$

$$\bar{y} = 50 \text{ mm} \qquad\qquad\qquad \textbf{Ans.}$$

Apply the parallel-axis theorem for each part and sum the result.

$$I_{x'} = \Sigma(\bar{I}_x + A d^2)$$

$$I_{x'} = \underline{\hspace{10cm}}$$

$$I_{x'} = 13.0(10^6) \text{ mm}^4 \qquad\qquad \textbf{Ans.}$$

10 - 6. The composite cross section for the column consists of two cover plates riveted to two channels. Determine the radius of gyration $k_{x'}$ with respect to the centrodial x' axis. Each channel has a cross - sectional area of $A_c = 7.50(10^3)$ mm^2 and a moment of inertia $(I_{x'})_c = 150(10^6)$ mm^4.

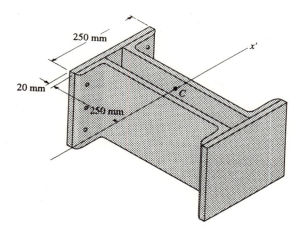

Solution

To determine $I_{x'}$ apply the parallel - axis theorem for each cover plate and include the moment of inertia of the channels.

$$I_{x'} = \Sigma(\bar{I}_{x'} + Ad^2) + 2\,(I_{x'})_c = \underline{\hspace{9cm}}$$

$$I_{x'} = 976.3(10^6) \text{ mm}^4$$

$$A = \underline{\hspace{7cm}}$$

$$A = 25.0(10^3) \text{ mm}^2$$

$$k_x = \sqrt{\frac{I_x}{A}} = \sqrt{\frac{976.3(10^6)}{25.0(10^3)}} = 198 \text{ mm} \qquad \textbf{Ans.}$$

Answers

1 - 1. $x = 1, y = 1, z = -1$

1 - 2.   a) $x = 4, -4$, b) $x = -1, 6$

1 - 3.   a) $50°$, b) $50°$, c) $50°$

1 - 4.   5

1 - 5.   a) 5, b) 7.07, c) 10

1 - 6. $\phi = 48.6°$, $\theta = 91.4°$, 9.33

1 - 7. $A = 28.5, B = 55, C = 88.3$

2 - 1. $\sqrt{(4)^2 + (10)^2 - 2(10)(4)\cos 120°}$

$$\frac{\sin\phi}{4} = \frac{\sin 120°}{12.5}$$

2 - 2. $\sqrt{(40)^2 + (60)^2 - 2(40)(60)\cos 105°}$

$$\frac{\sin\theta}{60} = \frac{\sin 105°}{80.3}$$

2 - 3. $\sqrt{(500)^2 + (600)^2 - 2(500)(600)\cos 20°}$

$$\frac{\sin\theta}{500} = \frac{\sin 20°}{214.9}$$

2 - 4. $$\frac{\sin 125°}{20} = \frac{\sin 35°}{F_1}$$

$$\frac{\sin 125°}{20} = \frac{\sin 20°}{F_2}$$

2 - 5. $\sqrt{(500)^2 + (400)^2 - 2(500)(400)\cos 30°}$

$$\frac{\sin\phi}{500} = \frac{\sin 30°}{252.2}$$

2 - 6. $$\cos\theta = \frac{400}{500}$$

$$\sqrt{(500)^2 - (400)^2}$$

2 - 7. $-2\cos 60°$
    $-2\sin 60°$

2 - 8. $-700\cos 30°$
    $700\sin 30°$

2 - 9. $75\cos 45° - 50\cos 30°$
    $75\sin 45° + 50\sin 30° - 50$

2 - 10. $100\cos 30° - 260(\frac{12}{13})$
    $100\sin 30° + 260(\frac{5}{13})$

2 - 11. $40\sin 30° - 60\cos 45°$
    $40\cos 30° + 60\sin 45°$

2 - 12. $20 = 30\cos\theta + 30\cos\phi$
    $0 = 30\sin\theta - 30\sin\phi$

2 - 13. $5\cos 60°\mathbf{i} + 5\cos 45°\mathbf{j} + 5\cos 60°\ \mathbf{k}$
    $-2\mathbf{j}$

2 - 14. $400\cos 120°\mathbf{i} + 400\cos 60°\mathbf{j} + 400\cos 45°\mathbf{k}$
    $600\cos 30°\cos 45°\mathbf{i} + 600\cos 30°\sin 45°\mathbf{j} - 600\sin 30°\mathbf{k}$

2 - 15. $-1.25\ \mathbf{i}$
    $-5\sin 30°\mathbf{j} + 5\cos 30°\mathbf{k}$
    $2(\frac{3}{5})\mathbf{j} + 2(\frac{4}{5})\mathbf{k}$
    $\sqrt{(-1.25)^2 + (-1.30)^2 + (5.93)^2}$
    $(\frac{-1.25}{6.20})$
    $(\frac{-1.30}{6.20})$
    $(\frac{5.93}{6.20})$

2-16. $200 \cos 135° \, \mathbf{i} + 200 \cos 60° \, \mathbf{j} + 200 \cos 60° \, \mathbf{k}$

$\{275 \, \mathbf{k}\}$

$0 = -141.4 + F_{2x}$

$0 = 100 + F_{2y}$

$275 = 100 + F_{2z}$

2-17. $50 \cos 30°$

$-50 \sin 30°$

$\dfrac{50}{\sin 60°}$

$57.74 \cos 60°$

2-18. $-200 \, \mathbf{j}$

$-400 \sin 30° \, \mathbf{i} + 400 \cos 30° \, \mathbf{j}$

2-19. $4\mathbf{i} + 8\mathbf{j} - 8\mathbf{k}$

$\sqrt{(4)^2 + (8)^2 + (-8)^2}$

$\left(\dfrac{4}{12}\right)$

$\left(\dfrac{8}{12}\right)$

$\left(\dfrac{-8}{12}\right)$

2-20. $x\mathbf{i} - 1\mathbf{j} - 2\mathbf{k}$

2-21. $\{ -4\mathbf{i} + 2\mathbf{j} + 4\mathbf{k}\}\,m$

$3\left(\dfrac{-4}{6}\mathbf{i} + \dfrac{2}{6}\mathbf{j} + \dfrac{4}{6}\mathbf{k}\right)$

2-22. $520\left(-\dfrac{10}{26}\mathbf{j} - \dfrac{24}{26}\mathbf{k}\right)$

$680\left(\dfrac{16}{34}\mathbf{i} + \dfrac{18}{34}\mathbf{j} - \dfrac{24}{34}\mathbf{k}\right)$

$560\left(\dfrac{-12}{28}\mathbf{i} + \dfrac{8}{28}\mathbf{j} - \dfrac{24}{28}\mathbf{k}\right)$

2 - 23.

$$\sqrt{(-120)^2 + (-90)^2 + (-80)^2}$$

$$\frac{-120}{170}, \; -\frac{90}{170}, \; -\frac{80}{170}$$

2 - 24.

$-3\mathbf{j}$

$2\mathbf{i} - 1\mathbf{j} - 2\mathbf{k}$

$0 + (-3) - 1) + 0 = 3$

$\cos^{-1}\left(\dfrac{3}{3(3)}\right)$

2 - 25.

$\{-1.5\mathbf{i} - 3\mathbf{j} - 4.5\mathbf{k}\}\text{m}$

$-1.5, \; -2, \; -6$

$(-1.5)\left(\dfrac{-1.5}{6.5}\right) + (-3)\left(\dfrac{-2}{6.5}\right) + (-4.5)\left(\dfrac{-6}{6.5}\right)$

2 - 26.

$\{6\mathbf{i} - 4\mathbf{j} - 12\mathbf{k}\}\text{m}$

$28\left(\dfrac{6}{14}\mathbf{i} - \dfrac{4}{14}\mathbf{j} - \dfrac{12}{14}\mathbf{k}\right)$

$-\mathbf{k}$

$0 + 0 + (-24)(-1)$

2 - 27.

$600\cos120°\mathbf{i} + 600\cos60°\mathbf{j} + 600\cos45°\mathbf{k}$

$(-300)(120) + (300)(90) + (424.3)(-80)$

$\dfrac{-42\,941}{170(600)}$

$(120\mathbf{i} + 90\mathbf{j} - 80\mathbf{k}) \cdot (1\mathbf{j}) = 0 + (90)(1) + 0$

3 - 1.

$F_2\cos30° - F_1\cos - 6\cos20° = 0$

$F_1\sin30^0 + F_2\sin30° - 6\cos20° = 0$

3 - 2.

$F\cos\theta - 98.1\cos45° - 98.1\cos75° = 0$

$98.1\sin75° - 98.1\sin45° - F\sin\theta = 0$

3 - 3.

$-T\cos\phi + T\cos\theta = 0$

$2T\sin\theta - W = 0$

$5(1)$

$\sin^{-1}\dfrac{1.5}{2.5}$

3 - 4. $-98.1 \cos\theta + 98.1 \cos\phi = 0$

$2(98.1 \sin\theta) - W = 0$

3 - 5. $T_{BC} \cos30° - T_{BA} \cos60° = 0$

$T_{BA} \sin60° - T_{BC} \sin30° - 39.24 = 0$

$39.24 \cos30° + T_{CD} \cos30° = 0$

$39.24 \sin30° + T_{CD} \sin30° - F = 0$

3 - 6. $150 - 2T \sin\theta = 0$

$2(107.1) \cos44.4° - m(9.81) = 0$

3 - 7. $300 - R \sin45° = 0$

$F + R \cos45° - P \cos30° = 0$

$P \sin30° - 500 = 0$

3 - 8. $-R \, \mathbf{k}$

$- \; 600 \, \mathbf{k}$

$400 \, \mathbf{j}$

$-1, \; -3, \; 3$

$\cos^2 60° + \cos^2 45° + \cos^2 \gamma = 1$

$\cos60°, \; \cos45°, \; \cos60°$

3 - 9.
4	-6	-12
-6	-4	-12
-4	6	-12

3 - 10. $-0.333 \, F_B - 0.8 \, F_C + 0.333 \, F_D = 0$

$- \; 280 - 0.667 \, F_B + 0.667 \, F_D = 0$

$- \; 500 + 0.667 \, F_B + 0.6 F_C + 0.667 \, F_D = 0$

3 - 11. $0.385 \, F_{BA} + 0.707 \, F_{DA} - 0.6 \, F_{AC} = 0$

$0.923 \, F_{BA} - 0.707 \, F_{DA} = 0$

$0.8 \, F_{AC} - 196.2 = 0$

4-1. $200(2)$

$260(\dfrac{5}{13})(6)$

$4\cos30° (4) - 4\sin30^0(2)$

4-2. $400(4)$

$250(\dfrac{3}{5})(4)$

$500\sin30° (12)$

4-3. $(150)\dfrac{4}{5}(8\sin30°) + (150)\dfrac{3}{5}(8\cos30°)$

$500\cos45° (7) + 500\sin45° (9)$

4-4. $\dfrac{3}{4}$

$5(30)$

4-5.
4	4	-2
40	60	-20

4-6.
-4	2	3
-6	4	3

4-7.
0	-6	-1
200	-200	-100

6	-12	-4
200	-200	-100

4-8. $6j$

0	6	0
60	-40	-120

4-9. $2\sin30°i + 2\cos30°j$

$-22k$

i	j	k
1	1.732	0
0	0	-22

4-10. $0.707\mathbf{i} + 0.707\mathbf{j}$
  $-0.707\mathbf{i} - 0.707\mathbf{j}$
  $-2\mathbf{i} + 3\mathbf{j} + 2\mathbf{k}$
  $-6\mathbf{i} - 1\mathbf{j} + 2\mathbf{k}$
  Using the above, one possible solution is

0.707	0.707	0
−2	3	2
30	40	20

4-11. $6\cos60°\mathbf{i} + 6\cos120°\mathbf{j} + 6\cos45°\mathbf{k}$
  $0.707\mathbf{i} - 0.707\mathbf{j}$
  $0.707\mathbf{i} + 0.707\mathbf{j}$
  $2\mathbf{i}$
  $2\mathbf{j}$

0.707	−0.707	0
2	0	0
3	−3	4.243

4-12. $(50\cos60°)\sin45°$
  $17.68(0.9)$

4-13. $150(\frac{4}{5})(14) + 150(\frac{4}{5})(12)$

  $150(\frac{4}{5})(14) + 150(\frac{3}{5})(16)$

  $150(\frac{4}{5})(14) + 150(\frac{3}{5})(16)$

4-14. $4\sin30°(2) + 4\sin30°(2) + 4\cos30°(4)$
  $4\cos30°(4) + 4\sin30°(4)$
  $4\cos30°(4) + 4\sin30°(4)$

4-15. $4/(0.005)$
  $4/(0.03)$

4-16. $17\cos25° + 25\sin25°$
  $25\cos25° - 17\sin25°$

4-17. $-4\mathbf{i} - 12\mathbf{j} - 2\mathbf{k}$
  $3, -6, -6$

−4	−12	−2
200	−400	−400

4 - 18. $0.2\mathbf{i} + 0.3\mathbf{j}$

$125\mathbf{k}$

0.2	0.3	0
0	0	125

4 - 19. 15 kN

$15(\dfrac{4}{5})(4) - 15(\dfrac{3}{5})(2)$

4 - 20. 8 kN

$8000(\dfrac{3}{5})(0.180) - 8000(\dfrac{4}{5})(0.06)$

4 - 21. $10\cos30°$

$10\sin30° - 20$

$30 + 10\cos30°(6) - 20(2)$

4 - 22. $2\cos45° - 3\sin30°$

$2\sin45° - 3\cos30°$

$3\sin30°(4) + 3\cos30°(4) + 2\cos45°(3) - 2\sin45°(5) + 4(2)$

4 - 23. $40 - 60(\dfrac{3}{5})$

$- 60(\dfrac{4}{5}) - 30$

$60(\dfrac{3}{5})(5) + 60(\dfrac{4}{5})(0.4) - 40(2)$

4 - 24. $\{50\mathbf{i} - 20\mathbf{j} - 30\mathbf{k}\}$ N

1	3	2
50	-20	-30

4 - 25. $0, 3, -4$

2	2	-3
0	60	-80

4 - 26. $-1.20\mathbf{i}$

0	1.5	2.5
-1.20	0	0

4-27. 40 [4, 0, −3]

   0   5   3

   32   0   −24

4-28. 4, −6, −12

   0, 9, −12

   60i + 18j − 324k

   6i − 8j + 1.5k or

   2i − 2j + 13.5k

   −8i − 4j + 13k or

   −8i + 5j + 1k

4-29. 350 + 175 + 75

   −75 (0.5) − 175 (1.1) − 350 (2.35)

4-30. 0 = 4x + 4.5(3) - 4(3) − 3(3)

   0 = −4y + 4.5(4) + 4(3) − 3(4)

4-31. 25 N

   25d = 25(1.5) − 30(2)

4-32. 200(0.75) − F_C (0.75 cos θ)

   300(0.75) − F_C (0.75 sin θ)

4-33. $\frac{1}{2}$ (10) (9)

   (5) (9)

   −45 − 45

   −45(3) − 45(4.5)

4-34. $\int_0^{10} \frac{1}{2} x^3 \, dx$

   $\int_0^{10} x \, (\frac{1}{2} x^3) \, dx$

4-35. 75 − 3

   125

   75(1) + 125 (0.15) − 3(1.4)

4 - 36. $75(0.5)$

$\frac{1}{2}(200)(3)$

$37.5 - 300 + wd = 0$

$37.5(\frac{1}{2})(0.5) - 300(2) + (wd)(3 - \frac{d}{2}) = 0$

5 - 1. $100\cos30° - 100\cos45° - B_x = 0$

$B_y - 100\sin30° - 100\sin45° = 0$

5 - 2. $-2(1) - 3(2) + N_B\cos30°(3) = 0$

$A_x - 3.079\sin30° = 0$

$A_y - 1 - 2 - 3 + 3.079\cos30° = 0$

5 - 3. $-(F_{CD}\sin40°)2 + 490.5(6) = 0$

$A_x - 2289\cos40° = 0$

$-A_y + 2289\sin40° - 490.5 = 0$

5 - 4. $-(F_{BC})_{max}(6\cos60°) + 58.86(40\cos60°) = 0$

$-392.4(10^{-3})(\sin28.68°)(6) + m(9.81)(50)\cos45° = 0$

5 - 5. $F'_B - 14\,715\sin30° = 0$

$-N_A(2) + 14\,715(\cos30°)(1.2) - 14\,715(\sin30°)(0.4) = 0$

$6174.6 - 14\,715\cos30° + N_B = 0$

$(6174.6/2)/58\,000$

$(6569.0/2)/65\,000$

5 - 6. $-300(2) + A_z(3) = 0$

$400(2) - B_z(4) = 0$

$200 + 200 + F_{DC} - 700 = 0$

5 - 7. $-F_{DC}(0.2) + 2(0.2) = 0$

$-A_y + 1 = 0$

$B_x(0.8) - 2(1.4) = 0$

$A_x - 3.5 + 2 = 0$

$-2(0.4) + B_z(0.8) + 1(0.2) = 0$

$A_z + 0.750 - 2 = 0$

5 - 8.

$$-800(4) - 600(4) - C_y(3) = 0$$
$$800(4) - A_z(4) = 0$$
$$A_y(4) - D_x(4) = 0$$
$$A_x + D_x = 0$$
$$A_y + C_y = 0$$
$$A_z + C_z - 1400 = 0$$

5 - 9.

$$0.7071T_{BD} - 0.6124\,T_{BC} + A_x = 0$$
$$-0.3536(T_{BC}) + A_y = 0$$
$$A_z - 0.7071T_{BD} - 0.7071T_{BC} - 1962 = 0$$
$$3.2\mathbf{k},\ 2.4\mathbf{j},\ -1962\mathbf{k}$$

6 - 1.

$$4 - F_{AB}\sin60° = 0$$
$$F_{AE} - F_{AB}\cos60° = 0$$
$$4.62\sin60° - F_{BE}\sin60° = 0$$
$$(F_{BE} + 4.62)\cos60° - F_{BC} = 0$$

6 - 2.

$$-400 + F_{CB}\left(\frac{1}{\sqrt{5}}\right) = 0$$

$$894.4\left(\frac{2}{\sqrt{5}}\right) - F_{CD} = 0$$

$$F_{BE}\sin53.13° - 800\sin63.43° = 0$$
$$F_{BA} - 894.4 - 800\cos63.43° - 894.4\cos53.13° = 0$$
$$F_{AE} - 1788.9\cos63.43° = 0$$

6 - 3.

$$3 - F_{CD}\sin22.62° = 0$$
$$-F_{CB} + 7.80\cos22.62° = 0$$
$$-7.80 + F_{DE} = 0$$

6 - 4.

$$100 - 50 - F_{CL} = 0$$
$$-100(6) + 50(3) + F_{ML}(4) = 0$$
$$F_{CD}(4) + 50(3) - 100(6) = 0$$

6 - 5.

$$F_{CM}\left(\frac{3}{5}\right) - 5 = 0$$
$$-F_{CB}(1.5) + 5(10) = 0$$
$$5(8) - F_{LM}(1.5) = 0$$

6 - 6. $-8(0.866 \cos 30°0 + F_{BF} \sin 60°$ (1) $= 0$

$F_{BC}(1.5 \tan 30°) + 8(1.5 - 0.866 \cos 30°) - 11(1.5) = 0$

$-11(1) + 8(0.5 \cos 60°) + F_{GF}(0.5) = 0$

6 - 7. $1.3375 - F_{CD}(\frac{3}{5}) = 0$

$-F_{GF} + 2.23(\frac{4}{5}) = 0$

6 - 8. $-800(2) - 400(4) + 1600(4) - F_{DF}\cos 45°(2) = 0$

$-800(2) - 400(4) + 1600(6) - F_{CE}\cos 45°(2) = 0$

$-2000 + F_{ED} - 2262.7\cos 45° + 4525.5\sin 45° = 0$

6 - 9. $4T - 800 = 0$

$200 - 5P = 0$

$-80(420) - 80(180) + 200(x) = 0$

6 - 10. $-600(1.5) + F_{BC}\sin 45°(3) = 0$

$-A_x + 424.26 \cos 45° = 0$

$A_y - 600 + 424.26 \sin 45° = 0$

6 - 11. $-800(1.9) + F_{BC}(\frac{3}{5})(4) = 0$

$-A_x + (633.3)(\frac{4}{5}) = 0$

$A_y - 800 + (633.3)(\frac{3}{5}) = 0$

6 - 12. $D_y(1) - 2.25(0.5) = 0$

$F_y - 1.125 = 0$

$-F_{BE}(1) + F_y(0.5) = 0$

$D_x - F_{BE} = 0$

6 - 13. $960(0.6) - T(\frac{5}{13})(1.2) = 0$

6 - 14. $-F_{FB}(\frac{3}{\sqrt{10}})(2) + 1226.25(3) = 0$

$E_x - 1938.9(\frac{1}{\sqrt{10}}) = 0$

$-1226.25 + 1938.9(\frac{3}{\sqrt{10}}) - E_y = 0$

$613.1(3) - F_{DB}\sin 45°(1) = 0$ 142

6 - 15. $-833.85(0.8) - 588.6(2) + B_y(3) = 0$

$A_y - 833.85 - 588.6 + 614.8 = 0$

$-C_y(1.5) - C_x(1) + 807.7(3) = 0$

$C_x(1) - C_y(1.5) - 614.8(3) = 0$

$-F_{DE} + 2133.7 = 0$

7 - 1. $5 - 4 - N_A = 0$

$5 - 5 + N_B = 0$

7 - 2. $-500 + 2 F(\frac{3}{5}) = 0$

$V_D - 416.7(\frac{4}{5}) = 0$

7 - 3. $-F_{CD}(\frac{4}{5})(0.06) - F_{CD}(\frac{3}{5})(0.4) + 20(0.16) = 0$

$B_x - 11.11(\frac{3}{5}) = 0$

$20 - 11.11(\frac{4}{5}) - B_y = 0$

$6.667 - V_E = 0$

$N_E - 11.11 = 0$

$-6.667(0.2) + M_E = 0$

7 - 4. $B_x(3) - 1.5(1.5) = 0$

$0.750 - V_F = 0$

0

$0.750(0.5) - M_F = 0$

8 - 1. $-0.3N_A + F - N_B = 0$

$N_A - 166.77 - 0.2N_B = 0$

$-F(1 \sin 50°) - 166.77(2 \cos 50°) - 0.2N_B(4 \cos 50°) + N_B(4 \sin 50°) = 0$

8 - 2. $-0.2N_B - 0.2N_A(\frac{4}{5}) + N_A(\frac{3}{5}) = 0$ $N_B = 2.2 N_A$

$N_A(\frac{4}{5}) + 0.2 N_A(\frac{3}{5}) + N_B - 500 - 176.58 = 0$

$176.58(1.75)(\frac{4}{5}) + 500(x)(\frac{4}{5}) - N_A(3) = 0$

8 - 3. $-P(1.5) + 1962(0.5) = 0$

$N_B - 1962 = 0$

$654\,\text{N}$

$0.4(1962)$

8 - 4. $-490.5(2) + N_B(3) = 0$

$N_C + P\sin30° - 327.0 = 0$

$P\cos30° - F_B - F_C = 0$

$-P\cos30°(0.3) + F_B(0.7) = 0$

$0.8(195)$

8 - 5. $P - N_A - F_C - 735.75\sin30° = 0$

$N_C + F_A - 735.75\cos30° = 0$

$-F_C(0.15) + F_A(0.15) = 0$

$N_A - F_B - 735.75\sin30° = 0$

$N_B - F_A - 735.75\cos30° = 0$

$F_A(0.15) - F_B(0.15) = 0$

$0.25(794.8)$

$0.4(479.5)$

9 - 1. $y\,dx = (1-x^2)\,dx$

x

$\dfrac{1}{2}y = \dfrac{1}{2}(1-x^2)$

$x\,dy = (1-y)^{\frac{1}{2}}\,dy$

$\dfrac{1}{2}x = \dfrac{1}{2}(1-y)^{\frac{1}{2}}$

y

9 - 2. $y\,dx = x^2\,dx$

x

$\dfrac{y}{2} = \dfrac{1}{2}x^2$

$(1-x)\,dy = (1-y^{\frac{1}{2}})\,dy$

$\dfrac{1+x}{2} = \dfrac{1+y^{\frac{1}{2}}}{2}$

y

9 - 3.

$$(y_2 - y_1)\, dx = (\sqrt{8}\, x^{1/2} - 2x)\, dx$$

x

$$\frac{1}{2}(y_2 + y_1) = \frac{1}{2}(\sqrt{8}\, x^{1/2} + 2x)$$

$$(x_1 - x_2)\, dy = \left(\frac{y}{2} - \frac{y^2}{8}\right) dy$$

$$\frac{x_1 + x_2}{2} = \frac{1}{2}\left(\frac{y}{2} + \frac{y^2}{8}\right)$$

y

9 - 4.

$$\pi y^2\, dz = \pi\,(0.4z)\, dz$$

z

9 - 5.

$$\pi\,(z^2)\, dy = 6.25\, y^4\, dy$$

y

9 - 6.

$$\frac{0(3) + (1.5)(3) + (1.5)(3\sqrt{2})}{3 + 3 + 3\sqrt{2}}$$

$$\frac{(1.5)(3) + 0(3) + 1.5(3\sqrt{2})}{3 + 3 + 3\sqrt{2}}$$

9 - 7.

$$\frac{25(\pi)(25)^2 + (50 + 55)15(110) + (50 + 110 + \frac{35}{2})\pi(\frac{35}{2})^2}{(\pi)(25)^2 + 15(110) + \pi(\frac{35}{2})^2}$$

9 - 8.

$$\frac{25(50)(350) + 2[(50 + 100)(50)(200)]}{(50)(350) + (2)(50)(200)}$$

9 - 9.

$$\frac{\pi}{4}(2.25)\,(0.5)(0.2)$$

9 - 10.

$$\frac{\pi}{4}[2.25(0.5) + 2.25(0.5) + 2(0.2) + 2.5(0.2)]$$

9 - 11.

$$(2\pi)(2)(4) + 2\pi(1)\left(\sqrt{(2)^2 + (1.5)^2}\right)$$

$$2\pi(1)(2)(4) + 2\pi\left(\frac{2}{3}\right)\left(\frac{1}{2}\right)(1.5)(2)$$

10 - 1. $35(10^6) = I_{DD} + 12.0\,(10^3)(40)^2$

$15.8(10^6) + 12.0(10^3)(10)^2$

10 - 2. $y^2\,2x\,dy = 40\,y^{\frac{3}{2}}\,dy$

10 - 3. $0.1(1600 - x^2)\,dx$

$x^2(0.1)(1600x^2 - x^4)\,dx$

10 - 4. $y^2(2x)\,dy = y^2(2\sqrt{200}\,y^{1/2})\,dy$

$2x\,dy = 2\sqrt{200}\,y^{1/2}\,dy$

$914(10^6) = \bar{I}_{x'} + 53.33\,(10^3)\,(120)^2$

10 - 5. $\bar{y} = \dfrac{20(40)(100) + 100(120)(20)}{40(100) + 120(20)}$

$I_x = \dfrac{1}{12}(100)\,(40^3) + (100)(40)(50 - 20)^2 + \dfrac{1}{12}(20)\,(120)^3 + (20)(120)(100 - 50)^2$

10 - 6. $2[\dfrac{1}{12}(250)(20)^3 + (250)(20)(250 + 10)^2] + 2[150(10^6)]$

$2(250)(20) + 2(7.50)(10^3)$